X-Ray Diffraction Imaging
Imaging
Technology and Applications

Devices, Circuits, and Systems

Series Editor
Krzysztof Iniewski
Emerging Technologies CMOS Inc.
Vancouver, British Columbia, Canada

PUBLISHED TITLES:

PUBLISHED TITLES:

PUBLISHED TITLES:

PUBLISHED TITLES:

FORTHCOMING TITLES

X-Ray Diffraction Imaging
Imaging
Technology and Applications

Edited by
Joel Greenberg
Managing Editor
Krzysztof Iniewski

CRC Press
Taylor & Francis Group
Boca Raton London New York

CRC Press is an imprint of the
Taylor & Francis Group, an **informa** business

CRC Press
Taylor & Francis Group
6000 Broken Sound Parkway NW, Suite 300
Boca Raton, FL 33487-2742

First issued in paperback 2021

© 2019 by Taylor & Francis Group, LLC
CRC Press is an imprint of Taylor & Francis Group, an Informa business

No claim to original U.S. Government works

ISBN-13: 978-1-03-209427-4 (pbk)
ISBN-13: 978-1-4987-8361-3 (hbk)

Library of Congress Cataloging-in-Publication Data

Names: Greenberg, Joel (Joel Alter), author. | Iniewski, Krzysztof, 1960- author.
Title: X-ray diffraction imaging : technology and applications / Joel
Greenberg and Krzysztof Iniewski.
Description: Boca Raton : Taylor & Francis, CRC Press, 2018. | Series: Taylor
and Francis series in devices, circuits, & systems | Includes
bibliographical references.
Identifiers: LCCN 2018027858 | ISBN 9781498783613 (hardback : alk. paper)
Subjects: LCSH: X-ray diffraction imaging. | Radiography, Industrial. |
Detectors—Equipment and supplies. | Refractometers.
Classification: LCC TA417.25 .G74 2018 | DDC 620.1/1272—dc23
LC record available at https://lccn.loc.gov/2018027858

Visit the Taylor & Francis Web site at
http://www.taylorandfrancis.com

and the CRC Press Web site at
http://www.crcpress.com

Contents

Contents

Series Editor

Krzysztof (Kris) Iniewski is managing R&D at Redlen Technologies Inc., a startup company in Vancouver, Canada. Redlen's revolutionary production process for advanced semiconductor materials enables a new generation of more accurate, all digital, radiation-based imaging solutions. Kris is also a founder of ET CMOS Inc. (http://www.etcmos.com), an organization of high-tech events covering communications, microsystems, optoelectronics, and sensors. In his career, Dr. Iniewski held numerous faculty and management positions at the University of Toronto (Toronto, Canada), the University of Alberta (Edmonton, Canada), Simon Fraser University (Burnaby, Canada), and PMC-Sierra Inc. (Vancouver, Canada). He has published more than 100 research papers in international journals and conferences. He holds 18 international patents granted in the United States, Canada, France, Germany, and Japan. He is a frequently invited speaker and has consulted for multiple organizations internationally. He has written and edited several books for CRC Press (Taylor & Francis Group), Cambridge University Press, IEEE Press, Wiley, McGraw-Hill, Artech House, and Springer. His personal goal is to contribute to healthy living and sustainability through innovative engineering solutions. In his leisure time, Kris can be found hiking, sailing, skiing, or biking in beautiful British Columbia. He can be reached at kris.iniewski@gmail.com.

Editor

Joel A. Greenberg is currently on faculty in the Electrical and Computer Engineering Department and the Graduate Medical Physics Program at Duke University, as well as a member of the Fitzpatrick Institute for Photonics. His current research focuses on computational sensing and its application to security, medical, and industrial imaging and detection, which involves a balanced collaboration with academia, industry, and government partners. Joel received his B.S.E. in Mechanical and Aerospace Engineering from Princeton University in 2005 and his Ph.D. in Physics from Duke University in 2012. He has published over 60 papers in the areas of nonlinear optics, cold atom physics, compressed sensing, and X-ray imaging, and holds patents in the space of X-ray diffraction imaging. He can be reached at joel.greenberg@duke.edu.

Contributors

Paul Evans
School of Science and Technology
Nottingham Trent University
Nottingham, United Kingdom

Joel A. Greenberg
Department of Electrical and
    Computer Engineering
Duke University
Durham, North Carolina

Adam Grosser
Redlen Technologies Corporate
Saanichton, British Columbia,
    Canada

Krzysztof Iniewski
Redlen Technologies Corporate
Saanichton, British Columbia,
    Canada

Dirk Kosciesza
Smiths Detection Germany
    GmbH
Wiesbaden, Germany

Manu N. Lakshmanan
Dept. of Radiology & Imaging
    Sciences
National Institutes of Health,
    Clinical Center
Bethesda, Maryland

Shuo Pang
CREOL-College of Optics and
    Photonics
University of Central Florida
Orlando, Florida

Keith Rogers
Professor of Materials/Medical
    Science, Cranfield Forensic
    Institute
Cranfield University
Bedford, United Kingdom

Chris Siu
Department of Electrical and
    Computer Engineering
    Technology
British Columbia Institute of
    Technology (BCIT)
Vancouver, British Columbia,
    Canada

Scott D. Wolter
Department of Physics
Elon University
Elon, North Carolina

Zheyuan Zhu
CREOL-College of Optics and
    Photonics
University of Central Florida
Orlando, Florida

Introduction to X-Ray Diffraction Imaging

X-rays have been used for an array of sensing, imaging, and detection tasks for well over 100 years. The birth of radiographic imaging followed mere months after Röntgen's "discovery" of the X-ray in 1895, and X-ray diffraction (XRD) was already in use for aiding in the determination of crystalline structure by 1912. The pioneering work of Max von Laue and the father–son team of W. L. and W. H. Bragg provided both demonstrations and interpretations of the physics behind XRD and have enabled a variety of refinements, generalizations, and simplifications of the method. As a result, XRD is now a quantitative, versatile, and ubiquitous tool that has been used across science and engineering.

The most common X-ray diffraction systems, however, measure only the XRD signal from a single location on the surface of a sample (i.e., they do not perform imaging). Thus, despite the commercial and scientific success of X-ray diffraction in general, its implementation as a non-imaging modality limits its applicability to the analysis of thin samples and/or surfaces only. As a consequence, many industries have historically not been able to use XRD for performing meaningful material characterization; instead, they have had to settle for transmission-based X-ray imaging, which provides the required spatial information but insufficient material specificity. The inadequacy of this approach is particularly apparent in cases in which both the shape and composition of a sample must be analyzed or in which the material of interest is concealed or otherwise obfuscated. Important examples of these scenarios arise in the diagnosis and detection of cancer in medical imaging and the detection of explosives and/or contraband items for security applications.

Various schemes for transforming XRD into an imaging modality have been proposed, dating back to the late 1980s. While these approaches were successful in demonstrating the capacity for realizing spatially resolved XRD analysis, they were prohibitively slow, expensive, and complicated, typically requiring some combination of a synchrotron X-ray source, cryogenically cooled, high-purity Germanium detectors, and minimal computational resources. As a result, X-ray diffraction imaging was regarded as an interesting prospect but went several decades without significant advancement or adoption for commercial application.

Over the last decade, though, two key technological innovations have revolutionized X-ray diffraction imaging. The first centers around the role and capabilities of computers in imaging systems today. The incredible growth of CPU and GPU performance in combination with theoretical insights in information and sensing theory have led to a paradigm shift in the approach to imaging system design in general. Alternative physical measurement architectures combined with novel post-processing algorithms allow for smarter, faster, cheaper, and more optimized imaging and detection systems. As an example of a computationally centric X-ray architecture, Chapter 1 of this book discusses a coded aperture approach to XRD imaging that leverages compressed sensing and machine learning techniques for improved performance. Similarly, Chapter 6 describes the newly developed XRD focal construct technology for performing fast, robust, depth-resolved XRD imaging. Finally, Chapter 7 introduces new measurement and data processing approaches that enable high-resolution XRD imaging with reduced scan times and dose, as well as methods to efficiently image textured, crystalline materials. Together, these techniques allow one to extract more information about a sample given a fixed set of resources, thus allowing XRD imaging to be realized at lower dose, shorter acquisition times, and lower cost, using inexpensive, off-the-shelf components.

The second technological innovation that has enabled practical XRD imaging is the development of efficient, multi-pixel spectroscopic detector arrays with shot-noise-limited performance. These components build on critical advances in the growth of high-quality, high-yield semiconductor materials, such as CdTe and CZT, as well as the development of low-noise electronics and attachment schemes. Such detectors are critical to XRD imaging, as the measurable XRD signal is generally weak, diffuse, and both energy- and angle-dependent; high-quality imaging therefore requires high signal-to-noise ratio measurements of the X-ray signal over each of these dimensions. The topics of semiconductor detector development and the associated integrated circuits are discussed in Chapters 2 and 3, respectively. Chapter 8 extends this discussion in the context of airport security systems by describing the specifications and design tradeoffs for such detectors as part of the inverse fan beam XRD imaging architecture.

Having established the tools and methods that have given rise to the recent renaissance in XRD imaging, this book continues by exploring a range of areas in which XRD imaging is currently having or could have a significant impact. Chapter 4 discusses applications of XRD imaging in medicine, including topics such as cancer detection, analysis of brain tissue, and bone mineral analysis. Chapter 5 discusses broadly the impacts and opportunities for XRD imaging in materials science, with a particular focus on processing–structure–property relationships and implications to security applications. In addition, Chapters 1, 6, 7, and 8 touch on other uses of XRD imaging, including explosives detection in aviation security, inspection of mail and cargo, process control in various industries, environmental monitoring, and food inspection.

This is the first book to focus on XRD imaging and is well-timed to capture both the remarkable recent progress and potential impact of this new and exciting field. All of the discussions in this book are based on the most state-of-the-art methods and results in the field and are written by leading international researchers who are performing groundbreaking work. Despite the cutting-edge nature of the content, the book is written to provide a self-contained historical and technical introduction to the topic so that experts and newcomers alike can understand and appreciate this powerful technology.

Thank you to all of our collaborators and colleagues that contributed to the content of this book. We hope that the readers will be as excited about the prospects for XRD imaging as we are, and that this book serves as a teaching tool and reference for the next generation of X-ray diffraction imaging scientists.

Joel Greenberg
Duke University
Dept. of Electrical and Computer Engineering

Krzysztof Iniewski
Redlen Technologies

MATLAB® is a registered trademark of The MathWorks, Inc. For product information, please contact:

The MathWorks, Inc.
3 Apple Hill Drive
Natick, MA 01760-2098 USA
Tel: 508-647-7000
Fax: 508-647-7001
E-mail: info@mathworks.com
Web: www.mathworks.com

1

Coded Aperture X-Ray Diffraction Tomography

Joel A. Greenberg

Duke University

CONTENTS

1.1 Introduction

X-ray photons, with wavelengths of 0.01 to 10 nm, reside in a unique place within the electromagnetic spectrum, in that they can investigate samples at two disparate length scales. On the one hand, X-rays can pass through centimeters of most materials and, in so doing, map out an object's properties at this macroscopic length scale. On the other hand, the fact that X-ray wavelengths are comparable to the atomic and/or molecular spacing of different materials means that one can determine the microscopic structure of a material. The interaction that makes the structural analysis possible, known as coherent scatter, underlies the mechanism of X-ray diffraction (XRD). X-ray diffraction is a well-established technique that represents a gold standard in materials analysis. Taken together, the capacity for X-rays to both penetrate the entirety and interact at the microscopic scale with a target object make possible non-destructive, volumetric imaging of the molecular structure of objects that are not transparent at conventional optical wavelengths. We refer to this imaging method as X-ray diffraction tomography (XRDT).

X-ray diffraction tomography has been explored since the 1980s as a method to provide spatially resolved information about the molecular structure of materials.[1,2] This technique has the potential to greatly complement existing tomographic imaging modalities, such as computed tomography (CT),[3] ultrasound,[4] and magnetic resonance imaging (MRI)[5] by providing orthogonal material information and enhanced contrast. However, traditional XRDT approaches have typically required excessively long scan times and/or large doses and specialized or expensive components.[6] As a result, XRDT has not yet emerged as a practical imaging modality, despite the extensive literature supporting the rich and relevant information contained within the XRD signal itself.[7]

A new approach to XRDT, known as coded aperture XRDT, was developed in the early 2010s and has been shown to reduce scan times by several orders of magnitude and requires only conventional, off-the-shelf X-ray components.[8,9] The technique uses a combination of physical coding (i.e., modulation of the XRD signal in real space) and computational processing to optimally extract information from the measured X-rays. This technique has been made possible by the exponential growth of computing power over recent decades, the development of advanced and compressive sensing algorithms, and the availability of simple and high-performing detectors and sources. The resulting CA-XRDT architecture is highly flexible and can be customized to a particular application and associated set of constraints. In this chapter, we introduce CA-XRDT in the context of previously developed XRDT methods, describe specific instantiations of the method (including practical considerations), and discuss areas in which the CA-XRDT approach provides distinct benefits. Examples of scenarios in which CA-XRDT is particularly well-suited to the problem space include bottle and luggage inspection for aviation or border security, the diagnosis and/or detection of cancer, food inspection, and environmental monitoring.

1.2 Overview of X-Ray Diffraction Tomography

1.2.1 Tomographic Imaging

Tomography involves the estimation of a higher dimensional object from the measurement of one or more lower dimensional projections of the object. The fact that the measurement dimension is usually smaller than the object's embedding dimension (i.e., dimension in which the object physically exists) implies that tomographic imaging involves multiplexed measurements.[10] Through a combination of measurement design and computational data processing techniques, one can invert (i.e., demultiplex) the acquired data and recover an estimate of the object. The mechanism for contrast in the image depends on the physics of the interaction between the X-rays and the sample which, in turn, strongly influences the overall measurement architecture.

A well-known example of tomographic imaging is X-ray computed tomography (CT), in which a multitude of transmission-based projection images of an object are acquired from different perspectives around the object (see Figure 1.1a). While each projection image multiplexes attenuation from all points along the direction of propagation, the measurement diversity includes many different multiplexing instantiations, which allows for an accurate and unique determination of the underlying object. For the case of single-energy CT, the contrast mechanism is based on photoelectric absorption, and the resulting image is a map of the electronic density throughout the measured object volume. Alternatively, one can perform multi-energy (or spectral) CT,[11] which enables one to recover a volumetric mapping of both the effective atomic number and electron density (or photoelectric and Compton interaction strengths) throughout the object.

While this scheme has the advantage that the measured X-ray signal is typically bright and highly focused, it has several drawbacks. For example, the requirement that one make many different measurements to estimate a single voxel renders imaging dynamic scenes challenging.[12] In addition,

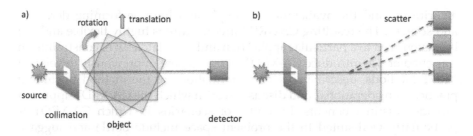

FIGURE 1.1
Schematic for tomographic measurement using (a) transmission and (b) scatter measurements. For transmission tomography, the requirement that each voxel be measured from multiple orientations implies that the rotation and/or translation of the object must occur. For scatter tomography, the angular spread of the scatter signal relegates the need for rotation and enables snapshot operation.

this leads to unnecessarily high-object exposure (including the exposure of unwanted regions of the sample), which is particularly undesirable for the case of medical imaging.[13] Finally, the contrast produced by attenuation alone may be insufficient for distinguishing particular features of interest.[14] In general, these contrast, dose, and speed-related issues stem from the fact that transmission-based imaging is sensitive only to X-rays that did not interact with the sample. As a result, all object-specific information is indirect, and the system makes poor use of the photons passing through the object because all directly interacting, or scattered, X-rays are blocked. While proposed methods, such as hyperspectral CT,[15] promise the potential for chemical sensitivity via transmission imaging, they still do not provide the molecular specificity often required in a variety of diagnosis- and/or analysis-focused tasks.

Rather than using the non-interacting, directly transmitted X-rays to perform tomography, one can instead produce an image of the object by measuring the scattered X-rays. While the underlying physics varies between scatter mechanisms, the fact that the scattered X-rays interact with the material means that their properties will have changed in a material-specific way. For example, the wave vector, energy, and/or polarization of the scattered photons represent new measurement degrees of freedom in scatter tomography that can provide additional contrast mechanisms (i.e., reveal additional information about the object of interest). Furthermore, these additional degrees of freedom result in a wide range of possible scattering tomography architectures (as will be discussed below in the context of XRDT), which open the door to novel tomographic capabilities. For example, the finite deflection of the scattered X-rays effectively lifts the degeneracy inherent in transmission tomography and makes snapshot (i.e., single measurement) scatter tomography possible (see Figure 1.1b). Alternatively, one can structure the illumination and measurement components to enable compressive tomography, in which the object dimensionality exceeds the measurement dimensionality.[10]

A challenge of scatter tomography, however, stems from the fact that the scatter signal is weak compared to the transmitted signal. This disparity in signal strengths arises from two distinct mechanisms: first, the scatter inter-action strength is small. As a result, the total scattered X-ray flux is usually only a few percent of the incident beam flux. Second, the scatter signal is spatially diffuse (i.e., the signal originating at any point within the object can be emitted into a large angular range) and decays with distance much more rapidly than the incident beam. While the discussion to this point has been general, the focus of the remainder of the chapter is on the specific case of XRDT and discusses methods for its successful implementation.

1.2.2 X-Ray Diffraction

X-ray diffraction is well-understood and can be interpreted in terms of coherent X-ray scatter from a phase-sensitive, multibody system. More spe-cifically, the electric field of an incident X-ray creates an oscillation of the electron charge distribution within the target object. This, in turn, causes the atoms to re-radiate X-rays at the same energy but spread out over a range of angles. This elastic scattering process, known as Rayleigh scatter, occurs for all atoms illuminated by the incident X-ray beam. For electrons located within the coherence volume of X-ray beam, the scattered fields have a well-defined phase relationship and can therefore interfere with one another. The resulting scattered X-ray distribution depends sensitively on the structure of the illuminated material and can therefore be used for material analysis (as discussed in more detail below).

The probability for an X-ray with energy E to undergo coherent scatter into a solid angled Ω is given by the differential scatter cross section

$$\frac{d\sigma_{coh}(E)}{d\Omega} = \frac{r_e^2}{2}\left[1 + \cos^2(\theta)\right] f(q) \tag{1.1}$$

where r_e is the classical electron radius and θ is the X-ray deflection angle (relative to the incident wave vector, see Figure 1.2a). The function $f(q)$ is the square of the coherent scatter form factor (or structure factor), which mod-ifies the free electron Thomson cross section by taking into account both intra-atomic and interatomic interference effects. This form factor is related to the Fourier transform of the electron density of an object and is therefore unique to each material. Furthermore, because it combines both chemical and structural information, $f(q)$ provides molecular specificity that allows one to go beyond standard chemical identification (as in fluorescence[16] or hyperspectral transmission imaging[15]). For example, diamond and graphite can be easily distinguished via XRD, despite having identical chemical com-positions. In this way, XRD provides a fingerprint that is material-specific and enables one to determine the molecular structure of a material and/or identify the type and class of an unknown substance. Figure 1.3a shows

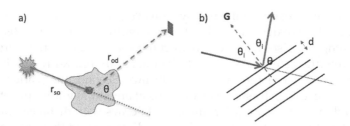

FIGURE 1.2
(a) Schematic for the basic geometry and key variables describing XRD. An incident X-ray along the line connecting the source and object scatter location, r_{so}, gives rise to a scattered X-ray along the line connecting the scatter location and detector, r_{od}. The deflection angle is θ. (b) For a crystalline lattice, the extra condition that the lattice grating vector bisect the incident and scattered X-ray wave vector must also hold.

example XRD form factors as a function of momentum transfer for a variety of materials; each material has a uniquely shaped $f(q)$ curve, as described by the locations, widths, and amplitudes of the different peaks.

It is instructive to consider first the case of a crystalline material, in which the atoms or molecules composing the material are arranged in a structured, long-range periodic lattice. In the Bragg interpretation of XRD, one can view each atom in this lattice as first coherently scattering the X-rays according to Eq. (1.1). The field produced from each scatterer coherently adds with that from every other, leading to constructive and destructive interference. Because of the crystal's long-range order and well-defined structure, there are only the particular directions in which constructive interference occurs and XRD is "allowed" (i.e., a macroscopic signal is measured). For a crystal with a lattice separation of d, the allowed scatter must satisfy Bragg's law (here shown in scalar form)

$$q \equiv \frac{1}{2d} = \frac{E}{hc} \sin\left(\frac{\theta}{2}\right), \tag{1.2}$$

where h is Plank's constant and c is the speed of light. The momentum transfer q is directly related to the momentum imparted to the scattered photon and is inversely related to the lattice spacing. From Eq. (1.2), one can see that a crystal with well-defined separations of its interatomic planes leads to narrow, well-defined "allowed" scattered angles at each incident energy, and the intensity of the scatter at these angles is given by the form factor $f(q)$. As an example of a well-studied material, Figure 1.3a shows the form factor for aluminum powder, which has a face-centered cubic (FCC) crystalline lattice.

While Bragg's law is necessary for describing the conditions for XRD to occur, it is not sufficient. In addition to arranging for measurement of the correct deflection angle and energy, X-ray diffraction from a highly crystalline

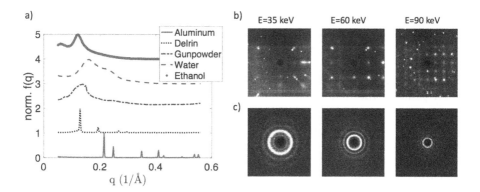

FIGURE 1.3
(a) Example form factors for aluminum, Delrin, gunpowder, water, and ethanol (bottom to top). Note that the crystalline materials give rise to sharply peaked form factors, while the fine amorphous or liquid materials give rise to broad yet unique peaks. (b) and (c) show the full scatter data over a range of deflection angles and multiple energies (35, 60, and 90 keV for the left, middle, and right columns, respectively) for a high purity crystal of silicon and for Delrin sheet, respectively. The silicon crystal produces patterned, discrete spots, whereas the amorphous Delrin gives symmetric Debye cones at all energies.

sample will only occur if the lattice grating vector G is co-planar with and bisects the incoming and outgoing X-ray wave vectors (as given by the Laue conditions and shown in Figure 1.2b). Thus, the resulting scatter pattern as a function of angle and energy depends both on how one chooses to illuminate the sample as well as how the scatter is measured. As an example, Figure 1.3b shows the scatter pattern obtained for a highly uniform Si wafer across a range of scatter angles and energies (left to right). The scatter signal is composed of discrete spots at particular angles and energies, as expected for a highly crystalline sample. The structure of the spots is particularly interesting in this case, as the Fourier transform relationship between the electron density and scatter distribution is readily apparent in their lattice-like distribution. For such samples, it is therefore not sufficient to simply measure the scatter pattern as a function of deflection angle θ and energy; characterizing the sample instead requires determining the full orientation distribution function (ODF) by making multiple measurements of the sample in different orientations relative to the beam.[17]

While single crystal analysis is conceptually straightforward and often is the method of choice for determining detailed crystal structure, the orientation dependence of the measured signal creates variability that makes material identification more difficult. Many applications instead involve first preparing the sample by crushing or grinding it into a very fine crystalline powder. For this case, one can view the sample as a polycrystalline ensemble, where each microcrystal is randomly oriented relative to the incident beam direction. The resulting diffraction signal, therefore, includes scatter from all possible lattice spacings consistent with Bragg's law (albeit with potentially different relative intensities, as compared to the case of a single crystal). Each lattice spacing creates a diffracted signal that is emitted into an azimuthally symmetric ring originating at the scatter origin (known as *Debye cones*, as shown in Figure 1.3c). Furthermore, these rings exist independently for each d spacing at every X-ray energy, with Bragg's law accurately relating the deflection angle and X-ray energy. Thus, while powder samples can produce very narrow peaks for highly crystalline materials, the scatter is more symmetric and spread out of angle and energy. As will be discussed in the next section, the spatio-spectral distribution of the signal is important when designing experimental configurations to adequately measure the signal.

We next consider non-crystalline materials, which include polymers lacking long-range order, glasses, and liquids. While such materials do not contain any long-range order, there does exist short range, local order due to the shape and interactions of the constituent atoms and molecules. Rather than having well-defined lattice spacings, amorphous materials are instead described by a radial distribution function that characterizes the probability of different interatomic (or intermolecular) separations being present in the material.[18] These radial distribution functions can contain peaks (which correspond to peaks in the corresponding XRD signal) but are generally much broader and usually contain only one or two local maxima. Examples of the form factors for materials in this class are shown in Figure 1.3a: Delrin has sparse, sharp peaks due to the organization of the long polymer chain, whereas water and ethanol have broad peaks due to the minimal degree of molecular organization.[19] Just as in the case of the polycrystalline samples discussed above, amorphous materials' lack of crystalline orientation results in the XRD signal being well-described in terms of azimuthally symmetric Debye cones. Figure 1.3c shows an example of the scatter signal obtained for a rod of Delrin, which produces symmetric rings with angular extents at each energy that are well-described by Eq. (1.2).

The above examples of perfect crystals, ideal powders, and perfectly amorphous samples are helpful for pedagogical discussions; however, most real materials fall somewhere between these platonic extremes. Depending on the history of the material, environmental parameters, degree of crystallinity, crystal size,[20] defects, etc., the scatter signal can look different for samples with the same molecular composition and lattice structure. In this way, XRD can be used for a variety of applications, ranging from detailed mapping of the crystal structure to detecting defects, material quality control, species identification, strain analysis, etc.[21,22] While this sensitivity is useful for detailed characterization tasks in which the material is known a priori, it can make identifying the composition of an unknown material more challenging.[23] This latter task, however, is potentially still achievable due to the relative invariance of the scattered peak locations in momentum transfer space, which are largely independent of the parameters described above.

1.2.3 XRDT Architectures

Given the rich information present in the XRD signal, the next question becomes how to design a measurement system to determine $f(q)$. To address this issue, we first consider the case of a single, thin sample material placed at a known location. Equation (1.2) shows that there are two different (although not mutually exclusive) methods for experimentally measuring $f(q)$, which are referred to as angle-dispersive XRD (ADXRD) and energy-dispersive XRD (EDXRD). In ADXRD, one measures the scatter intensity over a range of deflection angles at a single X-ray energy so that each angle uniquely corresponds to the scatter intensity at a particular q value (see Figure 1.4a). This can be accomplished in practice by using a well-collimated incident beam with either a monoenergetic (or narrowband) spectrum or by using a single energy channel from an energy-resolving detector. Alternatively, EDXRD involves measuring the XRD signal intensity over a range of energies for a fixed deflection angle (see Figure 1.4b).[2] One can implement an EDXRD scheme by illuminating the sample with a well-collimated, broadband X-ray beam and detecting the scatter using a range of energy channels from a collimated, energy-sensitive detector. Here, each energy corresponds to the scatter intensity at a particular value of momentum transfer. While the majority of commercial diffractometers make use of an ADXRD configuration for prepared sample analysis because of the excellent achievable momentum transfer resolution, EDXRD has the advantage that it can be more compact and operate without moving parts.

While the previous schemes enable accurate determination of $f(q)$ for well-localized samples, they are generally insufficient for performing tomographic imaging. This can be understood by considering again Eq. (1.2). In order to associate a detected X-ray of known energy with a particular momentum transfer value, one needs to know the deflection angle. However,

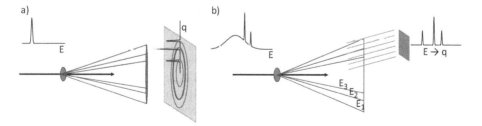

FIGURE 1.4
(a) Angle-dispersive XRD scheme, in which a spectrally narrowband illuminating beam gives rise to scatter from a known location. The scatter is measured at a range of different angles such that the form factor is directly related to the angle-dependent intensity of the scatter. (b) Energy-dispersive XRD scheme, in which a spectrally broadband illuminating beam gives rise to scatter from a potentially unknown location. A set of collimators enables the detector to only measure scatter from a single angle such that the form factor is directly related to the energy-dependent intensity of the scatter.

the determination of θ requires knowledge of the source, detector, and scatterer locations according to

$$\theta = \cos^{-1}\left(\frac{r_{so} \cdot \left(r_{sd} - r_{so} \right)}{|r_{so}||r_{sd} - r_{so}|} \right) \tag{1.3}$$

where r_{sd} and r_{so} are the vectors connecting the source to the detector and to the object, respectively. While the source and detector locations are often known, the location at which a particular X-ray underwent scattering is generally not known for a thick object. It is therefore necessary to determine the scatter origin for each detected X-ray in order to accurately associate it with a particular q. Based on Bragg's law, the momentum transfer resolution, Δq, ultimately depends on both the uncertainty in scatter angle and energy, $\Delta\theta$ and ΔE, respectively, according to

$$\frac{\Delta q}{q} = \sqrt{\left(\frac{\Delta E}{E} \right)^2 + \left(\frac{\Delta\theta}{\theta} \right)^2}. \tag{1.4}$$

As a general strategy, one typically seeks to approximately match the fractional uncertainty in energy and angle so as not to have one or the other dominate the system performance.

We note, however, that the need to determine the scatter origin for each individual X-ray can be relaxed in some situations and replaced by sufficient sampling of the distribution of scattered X-rays. For example, consider the case of an untextured object, in which each object position and momentum transfer value produces cones of scattered X-rays over a variety of energies. Through careful inspection of Eqs. (1.2) and (1.3), it is clear that the inherent nonlinearity in the mapping between q, E, and θ results in a unique spatio-spectral distribution of the scattered X-rays.[24] Thus, accurate measurement of the scattered X-ray intensity throughout angle AND energy (i.e., combining ADXRD and EDXRD) is sufficient to recover the full four-dimensional object $f(q,r)$. Said another way, the angle–energy correlations present in Eq. (1.2) allow one to infer the radiance of the scattered X-rays via a measurement of the energy spectrum at different locations in space; by comparing the measured scatter intensity distribution to a list of possible distributions, one can determine the characteristics of the scatterer (see Figure 1.5a).

While this spatio-spectral correlation approach is theoretically possible, there are two major challenges with it. First, the measured X-rays' energies must be known very accurately (typically less than 0.5% fractional energy resolution) over a large range of energies. Figure 1.5b shows an example of the spatio-spectral signal obtained along a linear array of energy-sensitive pixels (corresponding to the black plane in Figure 1.5a) for the two most similar points in the joint position and momentum transfer space. The ability

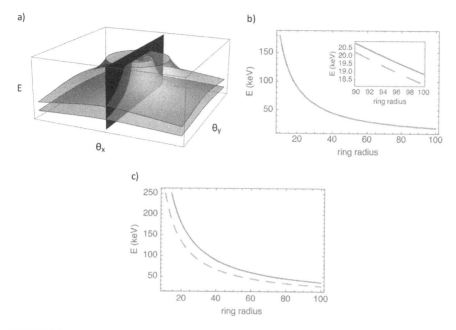

FIGURE 1.5

(a) Energy-angle curves for two different $f(q,r)$ objects, as given by Eq. (2 b) Energy-angle curve taken along a slice in the black plane shown in (a) for two different objects that are chosen to yield a scatter signal that is as close as possible to one another. The inset shows a zoomed-in version of the signals at large ring radii (i.e., deflection angles), showing a separation of only a few keV for this challenging case. (c) For the same objects as in (b), the energy-angle curves measured at a second plane with a different offset. The scatter signals are easily distinguishable in this second plane, despite their strong overlap in the first.

to distinguish these curves is generally beyond the state of the art for spectroscopic detectors (and physically impossible for certain energy ranges) although not unreasonable for tunable, monoenergetic X-ray sources. In addition, the presence of noise and multiplexing conspire to render such systems almost uninvertible for nearly any case of interest. If, however, one can record the scatter radiance (for example, by measuring the scatter signal at two distinct planes), the same two points in object space can be much more easily distinguished from one another (see Figure 1.5c).

It is therefore necessary to supplement the information content of the native signal by including additional elements that enable measurement of the radiance independent of any angle–energy correlations that may be present in the scatter signal. However, traditional detectors only record the irradiance (or intensity, independent of angle) of the incident X-ray signal; therefore, additional optical elements must be employed to provide the requisite angular information. While this is trivial in visible optics (e.g., using a lens), refractive elements are extremely lossy and challenging to fabricate at X-ray wavelengths.[25] As a result, the primary X-ray optical elements are absorptive in nature and act by simply blocking X-rays at certain locations to limit the volumes of the incident and scattered radiation. While this approach of structuring the spatial distribution of the incident and scattered X-rays is sufficient for recovering the information needed to realize XRDT, the overall photon efficiency can be quite low and traditionally must decrease in order for the imaging resolution to increase. Effective XRDT systems therefore require careful design of the illumination and detection systems in order to balance image fidelity and system throughput, subject to the constraints of the target application. In the rest of this section, we give an overview of the previously developed XRDT approaches in order to put the coded aperture XRDT approach in context.

1.2.3.1 Coherent Scatter Computed Tomography (CSCT)

A first approach to XRDT, known as coherent scatter computed tomography, was developed by Harding et al.[1] in the late 1980s. The approach is very analogous to transmission-based CT, in which one illuminates an object via different perspectives (using a combination of rotations and shifts of the object) with the exception that, in CSCT, one instead records the scattered X-rays (see Figure 1.6a). Each measurement is multiplexed over the volume of the illuminating beam and therefore requires the combination of multiple measurements (along with some variable degree of collimation) to demultiplex the signal and recover the underlying object, $f(q,r)$. While the initial concept and experimental demonstration were based on illumination with a pencil beam and discrete rotations and translations of the object (relative to the beam), a variety of variations have emerged over time. For example, one can alternatively use a fan beam to eliminate the need for translation but at the cost of reduced spatial resolution.[26,27] Also, depending on the incident beam geometry, one can choose to realize an ADXRD or EDXRD acquisition scheme by choosing the appropriate spatio-spectral dimensionality of the detector. As a complementary approach, we note that one can also employ detector-side collimation to better condition the system for inversion at the cost of overall collection efficiency.[28]

While not implemented in any commercial applications to date, CSCT has historically been the preferred method for XRDT analysis of small objects. Examples include medical applications as well as extensions to micro-CSCT for material characterization.[29] Compared to other methods, this approach typically has superior spatial and momentum transfer resolution (particularly in ADXRD mode) but is time intensive and requires both multiple, separate measurements and relative motion between the source and object. However, the ability to capture the whole two-dimensional scatter signal makes it possible to determine the ODF under sufficient measurement diversity (e.g., appropriate rotations and translations of the object).[17] As discussed in Chapter 7 of this book, recent innovations in the measurement and estimation space have led to more efficient, higher dimensional measurement

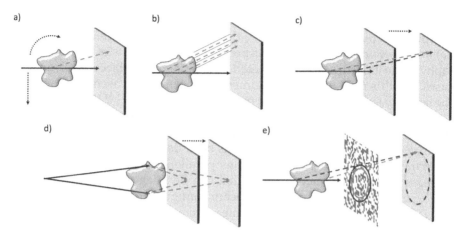

FIGURE 1.6

Schematics for various XRDT techniques, including: (a) coherent scatter computed tomography (CSCT), in which an object is rotated and/or translated in order to collect scatter from multiple orientations; (b) direct tomography (DT), in which collimators are used to map directly the scatter from each voxel to a distinct detector element; (c) kinetic depth effect X-ray diffraction (KDEXRD) imaging, in which a single detector is moved to multiple planes to collect scatter; (d) focal construct tomography (FCT), in which an annular beam illuminates a cone in the object that produces bright, focal spots at different planes that are unique to the location and composition of the object; and (e) coded aperture XRDT (CA-XRDT), in which a coded aperture is placed in the scatter path to uniquely modulate the scatter from each object point.

schemes that allow one to capture more information, limit the dose, and speed up the overall measurement.

1.2.3.2 Direct Tomography (DT)

While CSCT relies on a group testing measurement approach to overcome the high degree of intrinsic multiplexing, direct tomography[2] takes the opposite approach and seeks to avoid multiplexing altogether. This is accomplished by limiting the illumination volume and detector field of view (FOV) such that there is an isomorphic (i.e., direct) mapping between each object voxel and detector pixel (see Figure 1.6b). To realize this in practice, one typically employs narrow and precise collimation at the source and detector.[30,31] In order for each individual pixel (corresponding to only a single value of θ) to record the scatter intensity for an entire range of momentum transfer values, each pixel must be energy sensitive. Thus, direct tomography is inherently an EDXRD technique and, as such, relies on high-quality, energy-sensitive detectors.

Direct tomography has a number of distinct strengths. For example, it enables static, snapshot XRDT for one- and two-dimensional illumination volumes (e.g., pencil and fan beams), which may be critical when measurement resources are expensive and/or relative motion of the object, source, and detector are not possible. Direct tomography also obviates the need for computation post-processing, since the scatter signal is directly related to $f(q)$ and, therefore, is trivially interpretable. These aspects of direct tomography make it particularly well-suited to imaging large objects, such as suitcases. For this reason, several commercial systems have been developed in the security space to detect threat or contraband objects.[32-34] While the most common realization of DT systems involves an incident pencil beam to image $f(q)$ along a single line through the object, a multi-generational approach (similar to that in conventional CT) has been developed and implemented by Harding et al.[35] Higher generations parallelize the measurements and increase the object dimensionality that can be imaged per measurement. For example, fan and "inverse fan" beam configurations enable more rapid XRDT of a planar slice in a single measurement but require more detector elements as well as sources and collimation systems with increased complexity.[36]

While successfully employed in a variety of systems over the last three decades, DT faces two main challenges. First, high-quality imaging (i.e., good spatial and momentum transfer resolution) requires extremely narrow angular collimation at the detector. This need to block the majority of the scattered X-rays in order to obtain a well-conditioned signal, however, often results in extreme photon starvation and a correspondingly poor signal to noise (SNR) ratio.[37,38] Second, while the spectral resolution of the detectors is critical to the system's momentum transfer resolution, the required spectroscopic, multi-pixel detector arrays capable of operating over a broad range of hard X-ray energies are usually expensive and/or have insufficient spectroscopic performance to compete with traditional ADXRD systems (for more details on novel work in this space, see Chapters 2 and 8 in this book). Thus, while the spatial resolution is unaffected by these challenges, the system sensitivity is often low, and the recovered form factor is often extremely noisy for systems that do not employ ultrabright sources (such as synchrotrons).

1.2.3.3 Kinetic Depth Effect X-Ray Diffraction (KDEXRD) Imaging

The previously discussed CSCT and DT techniques are both sufficient for XRDT; however, they represent extremes in terms of the number of required measurements, degree of multiplexing, and need for computational post-processing. Kinetic depth effect X-ray diffraction imaging seeks to bridge these two approaches by simultaneously allowing multiplexing while requiring only a few discrete measurements.[39] Using a narrowband, pencil beam X-ray source, one records the ADXRD signal across a two-dimensional detector array with the detector placed at different depths from the object (see Figure 1.6c). By "tracking" the trajectory of different XRD rings as a function of propagation distance from the scatterer, one can realize a limited degree of angular sensitivity at the detector without the need for collimation. While this approach has been realized in proof of concept experiments, its utility is primarily limited to imaging thin, crystalline objects (i.e., materials with sparse, sharp features in $f(q)$). These constraints arise because the range over which the detector must be moved and the number of locations at which the detector must be placed increases rapidly as the target image resolution improves and the object size and degree of multiplexing increase due to the relatively poor system conditioning.

1.2.3.4 Focal Construct Tomography (FCT)

Focal construct tomography is a technique that effectively melds DT and KDEXRD to overcome some of their respective weaknesses. As described in further detail in Chapter 6 of this book, FCT involves illuminating an object with an annular beam.[40,41] Since XRD produces azimuthally symmetric, conical emission in the absence of texturing, the scatter resulting from each planar slice and specific momentum transfer value within the object produces a bright "focal" spot along the axis of the incident beam. The location of the focal spot, both in energy and range from the scatter origin, depends on the location of the scatterer and its momentum transfer values (see Figure 1.6d). Thus, by scanning a detector toward and away from the object (or using a fixed, energy-sensitive detector[42,43]) and noting the locations of the focal spots, one can estimate $f(q)$ along the incident beam direction. Because the intensity of the scatter at the focal spot is much larger for the correctly focused plane within the object than for other locations, FCT does not require significant collimation in order to isolate the signal of interest. As a result, it does not suffer the reduction in signal intensity present in DT. Furthermore, it relaxes the constraint of one-to-one imaging and therefore reduces the requirements on detector extent imposed by DT. Similarly, by structuring the incident beam in the manner described above, the system is much better conditioned than in KDEXRD. Thus, while FCT is an effective method for imaging a quasi-1D volume of an object, it does not scale straightforwardly to higher object dimensionality and therefore requires raster scanning in order to build up a complete volumetric image.

1.2.3.5 Coded Aperture X-Ray Diffraction Tomography (CA-XRDT)

A final approach to XRDT, and the focus of the remainder of this chapter, is a technique known as coded aperture X-ray diffraction tomography (CA-XRDT). At a high level, CA-XRDT can be viewed as a generalization and extension of the previously described methods, in which one combines physical coding (implemented via coded apertures) and computational inversion techniques to recover $f(q)$ throughout a subspace of the object. Depending on the properties and placement of the coded apertures, one can tune key system parameters, such as spatial resolution, photon throughput, degree of multiplexing, etc., in order to realize novel measurement configurations that can be optimized to particular tasks. As a simple example, we show in

Figure 1.6e a configuration very similar to the DT example in Figure 1.6b, in which a pencil beam illuminates an object and a multipixel detector records the scatter signal. However, rather than using collimation between the object and detectors, one can instead employ a high transmissioncoded aperture. The scatter signal is modulated by the coded aperture in a position-dependent way, thereby allowing for the recovery of $f(q,z)$ from a single measurement with orders of magnitude higher scatter signal intensities than the comparable DT approach, albeit with additional multiplexing. To further discuss the operational principles and advantages of CA-XRDT systems, the next sections describe the mathematical formalism, implementation-related details, and utility of the approach.

1.3 Coded Aperture X-Ray Diffraction Tomography (CA-XRDT)

As discussed in the previous sections, the central challenges in XRDT amount to (a) devising a method to determine the scatter origin, deflection angle, and energy of each scattered X-ray photon and (b) collecting enough X-rays from each voxel to faithfully estimate $f(q)$. Conventional approaches to XRDT suffer from inherent tradeoffs between the resolution and photon throughput (or, relatedly, scan time). For example, increased collimation in DT gives better spatial resolution but throws away a larger fraction of the measureable, scattered X-rays. Similarly, improving resolution in CSCT requires one to increase the number of sequential measurements obtained and/or increase the collimation, which correspondingly slows down the measurement.

Overcoming this tradeoff requires the ability to make better use of both the incident and scattered X-rays while simultaneously maintaining good image resolution. The key to realizing such a system is parallelization: by illuminating multiple voxels at once and simultaneously recording a large fraction of the scatter from each voxel, the overall photon efficiency of the system can be greatly increased. The trick, however, is to structure the incident illumination and generate scattered signals so that one can demultiplex the bright but complicated signal arriving at the detector and, in post-processing, determine the scattering properties of each voxel independently. This general scheme is shown in Figure 1.7a. This structuring can be implemented across space, time, energy, polarization, and angle, with the primary goal of creating a maximally distinct detected signal for each point in the object. While focusing optics would be a clear choice for constructing such a system, their high loss at X-ray energies typically overcomes any potential photon efficiency gains. Instead, we employ high-transmission coded apertures, which are thin, patterned opaque optical elements.[8] In contrast to collimators, coded apertures attenuate the X-ray signal at certain locations in the aperture plane but transmit X-rays at all angles. Thus, scatter from a broad range of locations can be visible to the detector at the same time. However, each point projects a unique shadow of the coded aperture onto the detector by virtue of the position-dependent parallax and/or magnification. By varying the properties of the coded apertures (discussed further below), one can control the degree of multiplexing and, relatedly, the imaging performance of the system. Furthermore, one can generalize this approach to include sequential measurements in which some combination of the source, detector, code, and object vary in time in order to provide additional structure and measurement diversity.[44]

Figure 1.7 shows a schematic example of a general coded aperture XRDT system. A first-coded aperture can be placed between the source and object to select certain illuminating rays. While this aperture effectively acts as a complicated collimating element for a single focal spot source, it allows one to identify the source of the incident ray when multiple focal spots are employed simultaneously. Each incident ray generates scattered X-rays throughout the volume of its intersection with the object. The scatter then passes through a second-coded aperture, which modulates the signal prior to its arrival at the detector. The role of this detector-side coded aperture is illustrated in Figure 1.7(b) and (c): two scatterers may generate a scatter signal that overlaps in space and energy, but the resulting modulation will be different. For two scatterers located at the same axial distance from the detector but shifted in the transverse direction, for example, the scatter overlaps different parts

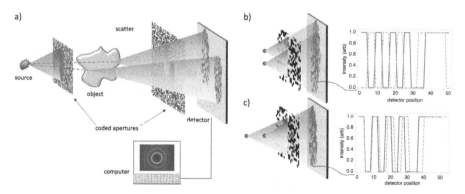

FIGURE 1.7

(a) Schematic of a CA-XRDT system, in which the conical radiation emitted from a conventional X-ray tube is modulated by a source-side coded aperture prior to illuminating the object. The generated scatter is then modulated by a detector-side coded aperture, which imprints the scatter origin onto the signal. A detector collects the scatter, and a computer processes the measurement. (b) Two scatterers at different lateral locations give rise to a modulated scatter signal that is shifted with respect to one another, whereas (c) two scatterers located at different axial locations give rise to signals that are modulated with different code magnifications (i.e., spatial frequencies).

of the coded aperture and, therefore, projects different pieces of the pattern onto the detector. In the region in which the scatter overlaps on the detector, one still obtains a relative shift of the projected code patterns. In contrast, for two scatterers separated axially, the projected code pattern is identical in structure but appears with different, position-dependent magnifications. Thus, for codes with orthogonality under shift and/or scale operations, the contributions from the different scatter locations can be determined independently. One can then use a final computational post-processing step to accomplish the demultiplexing required to estimate the underlying scatter properties of the object.

We note that the CA-XRDT approach explicitly assumes that the scatter signal is spatially or spectrally distributed and, therefore, has at least some extent in a dimension that can be coded. This is the case for amorphous, non-textured materials (such as liquids, powders, and many polymers), which produce Debye cones at a range of locations and energies. For the case of highly textured, crystalline materials, the XRD signal is typically localized at particular energies and angles. While some work has been done on understanding the impact of texturing on CA-XRDT,[23] mitigation and/or exploitation of this effect is still an area of active research.

Before continuing further with CA-XRDT, it is worth noting that elements of other XRDT systems are present as a subset of the coded aperture approach. For example, the annular beam shaping element in FCT is an example of a particular choice of source-side coded aperture, and such systems operate without the need for a secondary mask. Similarly, CA-XRDT is analogous also to KDEXRD, in that one can obtain radiance information about the XRD signal by measuring the scatter at different depths from the scatterer; however, CA-XRDT can realize such a scheme in a single measurement by using a detector-side-coded aperture, which effectively acts as a "virtual" secondary plane in which the scatter is sampled. Finally, CA-XRDT can be thought of as DT in the limit of poor collimation (i.e., collimators for which the angular acceptance is large) and for non-trivial collimator spatial structures. Thus, the coded aperture approach is clearly very broad and flexible and, most importantly, enables one to optimize the system for different sets of tasks and/or constraints.

1.3.1 Key CA-XRDT Components

The success of current CA-XRDT systems rests upon relatively recent improvements in the core components of the system. Advances in fast, efficient digital X-ray detectors, rapid improvements in computational power, and the development of new algorithms and data processing techniques are all necessary for practical implementation of CA-XRDT. In the following sections, the key components and considerations for CA-XRDT systems are discussed so that the methodology for targeted system design can be elucidated.

1.3.1.1 Coded Aperture

Coded apertures were first developed in 1949 by Golay[45] in order to overcome the tradeoff between resolution and throughput in the area of optical spectroscopy. It was shown that, by making multiplexed measurements and separating the signals after the fact, one could successfully trade physical measurement resources for computational resources, thereby speeding up the signal acquisition phase without sacrificing imaging performance. In fact, for the case when the measurement noise is not directly related to the signal amplitude (as in the case of Gaussian detector noise), one obtains an overall SNR improvement relative to making sequential, separate measurements. This multiplex advantage (or Felgatt's advantage[46]) leads to the creation of Hadamard spectroscopy[47] and clearly demonstrates the power of multiplexed measurement schemes. Since its original demonstration, coding in the visible optics domain has continued to expand and involve and, with the development of biased, nonlinear and/or decompressive estimation algorithms, allows one to straightforwardly overcome conventional space-time-spectral-polarization tradeoffs[48–51] and enable fast, snapshot acquisition of high-dimensional data sets. Furthermore, a variety of schemes to compressively and/or adaptively[10] sample and process the data have been developed and allow one to further improve the overall measurement efficiency in these systems. For example, coded aperture multiplexing can improve system sensitivity and SNR for compressible objects, even when photon noise (i.e., signal-dependent Poisson statistics) is dominant.[52]

Beyond enhancing hyperspectral imaging in the conventional optical regime, coded apertures were also applied early on to X-ray astronomy, where one seeks to image a scene of stars (i.e., bright, point-like objects) on a dark background. Previous systems had employed a pinhole camera for forming images; to overcome the intrinsic tradeoff between throughput and resolution as the size of the single pinhole increases, however, coded apertures structured as Fresnel zone plates[53] or multi-pinhole arrays[54,55] were employed. To recover sharp images of the stars with enhanced SNR (relative to the case of a single pinhole), one would then digitally correlate the recorded, multiplexed image with a complementary code.

In the CA-XRDT scheme, there are two primary locations that one can place the coded aperture —between the source and object and between the object and detector. The former coded aperture allows one to limit the illuminated volume in a controlled way. Imaging a full three-dimensional volume with a source-side code therefore requires that one employ relative motion between the source, coded aperture, or object (as discussed in section on "cone beam CA-XRDT" below) to ensure that no locations are missed. We note that, because the primary beam is so bright (compared to the scatter signal of interest), both primary and secondary "guard" coded apertures are typically used to eliminate unwanted scatter background from the primary coded aperture.

The detector-side coded apertures, on the other hand, act to reduce the degree of multiplexing (depending on the code transmission fraction) and imprint the scatter origin of a particular Debye ring. In order for the demultiplexing (or object estimation) process to be successful, it is important that the projection of the code pattern onto the detector be orthogonal between the distinct points one wants to resolve. The majority of codes are designed for performing two-dimensional imaging in a transverse plane and are therefore orthogonal under transverse shift (i.e., the correlation between the mask pattern and an in-plane, translated version of itself is very close to zero). For properly patterned codes, a shift of the projected code pattern of a single code feature of extent L_f is typically sufficient to separate neighboring points in the object space. Ideal codes not only have an autocorrelation that drops off quickly with shifts but also have minimal side lobes. The resulting transverse resolution is approximately equal to

$$\Delta r_{trans} = \frac{L_f (z_d - z_o)}{z_d - z_m} \tag{1.5}$$

where z_d, z_o, and z_m are the axial distances from the source to the detector, object, and mask, respectively.[56] The goal in designing such codes, then, is to choose (1) small code features and (2) maximize the ratio of the object-detector and mask-detector distances (so that small shifts in the object space are magnified upon propagation to the detector). Examples of common shift codes include Hadamard, uniformly redundant arrays (URAs)[57] or modified uniformly redundant arrays (MURAs),[58] and random codes, some of which are shown in Figure 1.8.

For distinguishing two points separated along the axial direction, it is critical that the code instead be orthogonal under scale transformations due to presence of the position-dependent magnification. The resulting axial resolution is approximately given as

$$\Delta r_{axial} = \frac{L_f (z_d - z_o)^2}{L_d (z_d - z_m)} \tag{1.6}$$

where L_d is the illuminated length of the detector. The quantity L_d/L_f is effectively the number of projected code features, which determines the ultimate resolution because one can distinguish smaller frequency changes (i.e., object position shifts) as one samples more of the total modulation. This scaling also suggests the use of small code features but implies an optimal mask location to realize the best axial spatial resolution over a target FOV. A variety of scale codes have been discussed,[59] but the simplest and most common are binarized, sinusoidal codes and low-length MURA codes. It is worth noting that one can combine scale and shift codes in multiple dimensions to satisfy the requirements of three-dimensional imaging, as in the

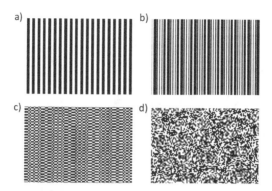

FIGURE 1.8

Examples of coded aperture patterns, including (a) a 1D periodic code, (b) a 1D MURA code, (c) a 2D periodic-MURA code, and (d) a 2D random code.

case of a two-dimensional code that is periodic along one dimension and a MURA code along the second (see Figure 1.8). Note that one can also use grayscale or "color" codes (i.e., using features with different amounts of energy-dependent attenuation).

While the resolution analysis above based on ideal-coded apertures (i.e., perfectly thin and attenuating) argues for small features, there are three practical caveats to this conclusion. First, high-contrast modulation requires code features with a non-zero thickness; as the thickness increases, the smallest allowed feature size consistent with a given aspect ratio (i.e., degree to which the coded aperture also acts as a collimator) increases. While the details depend on the choice of X-ray energies and material that the coded aperture is made of, a useful rule of thumb is that the smallest available feature size is about 4–5x the half value layer of the coded aperture material. For most applications, this imposes a limit of feature sizes on the order of a few hundred microns to a millimeter. The second practical limitation on the code feature size is that one must sample the modulated X-ray scatter sufficiently at the detector. Using Nyquist sampling as a good basis for estimation, this suggests that the smallest code feature should be on the order of the pixel size (although the geometry of the configuration can relax this constraint). Third, the code feature size should be larger than the focal spot size of the X-ray source in order to retain high-contrast and minimize penumbra-related effects.

There are several well-established techniques for fabricating high-quality coded apertures. Based on the energies of interest, one should first select a material with sufficient stopping power. In many cases, one can begin with a sheet of the target material and use conventional machining or etching techniques to remove the material at the desired locations.[60] For example, one might use computer aided drawing (CAD) software and a computer numerical control (CNC) machine to create patterns in a sheet of tungsten (see Figure 1.9a). While this approach gives clean features with high contrast

a)   b)

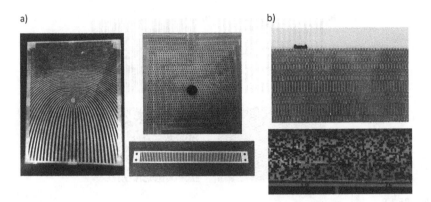

FIGURE 1.9
Coded apertures manufactured (a) by machining sheets or ingots of material and (b) by 3D printing the mask "negative" and filling in the interstitial places with a sufficiently attenuating powder (only zoomed in region of coded apertures shown.)

(i.e., the "open" regions contain only air), the drawbacks are that it is difficult to realize sharp, square features, the possible patterns cannot include floating features (i.e., regions of opaque material surrounded by air), and the maximum geometric thickness of the material severely limits the minimum realizable feature size. As an alternative, one can instead use additive manufacturing techniques such as three-dimensional printing (see Figure 1.9b). In this approach, one prints a negative of the target mask with a low-scatter and low-attenuation material. The empty sites are then filled in with a highly attenuating powder or slurry and then locked in place using a sealant (e.g., epoxy). The high resolution of today's three-dimensional printers allows for sharp features in arbitrary patterns but with reduced contrast due to the interstitial printer plastic. In the future, this issue may be resolved through the use of new three-dimensional printing materials that can be doped with higher atomic number particles to be used as the main attenuating elements.[61] Finally, it is worth noting that some attempts have been made at creating dynamic X-ray coded apertures by filling chambers with liquid metal solutions.[62]

1.3.1.2 X-Ray Source

The next critical element in any CA-XRDT system is the X-ray source. The key characteristics of the source include the size and number of focal spots, the source spectrum (both in space and energy), and power. As CA-XRDT results benefit from increased measurement diversity, the ideal source would be bright and tunable in time, space, and energy. While a synchrotron or related source provides the best performance, it is not feasible for use in most practical applications due to its size and cost. Special sources have also been created, including high-power multi-focal arrays,[63] but these are also expensive and require significant maintenance. Such sources introduce exciting possibilities when combined with coded aperture XRDT, but the focus of this section will be on the use of conventional Bremsstrahlung sources with a single focal spot and conical emission over a range of energies, as they are most readily available, economical, and well-established.

The focal spot size is important because it directly affects the achievable imaging resolution: the focal spot size sets a lower bound on the achievable Δr_{trans} as well as the smallest useful code feature size (which impacts Δr_{axial} as well). While small focal spots are preferred, the maximum tube power is proportional to the focal spot size: realizing high flux sources therefore requires larger focal spots (or more non-traditional approaches, such as liquid metal jet anode X-ray sources[64]). Thus, there is an intrinsic tradeoff between the achievable image resolution and scatter signal level. Depending on the target application, different specifications will lead to different choices in this trade space. For example, high-resolution imaging of tissue samples will be better served with a small focal spot, whereas hard X-ray imaging of thick (i.e., multi-cm) objects in baggage scanning is benefitted by higher tube powers. We note that we do not explicitly consider spatially coherent, microfocus sources in the following discussion.

Beyond the absolute intensity of the produced radiation, its distribution in energy and angle is also important. As discussed below, the choice of energy spectrum depends on whether an ADXRD or EDXRD (or both) scheme is used. Either spectral filtering at the source or the use of energy-sensitive detection can adequately constrain the range of contributing X-ray energies and, thereby, provide the information required to determine the associated momentum transfer of detected X-rays. In the case of filtering, one typically seeks to combine the proper choice of spectral filter material (which removes signal at unwanted energies) and tube anode material and source voltage (which generates intrinsically sharp and bright characteristic lines and shapes the Bremsstrahlung spectrum in the region of interest). In contrast, for the case of energy-sensitive detection, it is more useful to have a uniform and broadband spectrum to access a large range of momentum transfer values within the finite dynamic range of the detector. With regard to the angular distribution of the emitted radiation, some CA-XRDT architectures require and exploit a wide range of angular diversity,[65] whereas others employ conventional source-side collimation to create limited, well-defined illumination configurations (such as a pencil or fan beam). Thus, the proper choice of X-ray source is a flexible but important parameter in the overall system design.

1.3.1.3 Detectors

X-ray detectors have a long history and broad range of configurations and implementations. Chapters 2, 3, and 8 of this book discuss this topic in depth;

the focus in this section is instead on the key detector parameters as they relate to CA-XRDT performance. In general, we divide detectors into three main categories: energy integrating, photon counting, and energy sensitive (or spectroscopic). Energy-integrating detectors usually involve scintillator crystals to indirectly convert incident X-ray photons into an analog current that is proportional to the total energy deposited in the detector material. This confounds measurement of the number and energy of the incident X-ray photons as well as the intrinsic detector quantum efficiency (DQE) and makes it nearly impossible to separate photon noise from other detector-related sources of noise (such as dark or read noise). In contrast, photon-counting detectors register the arrival of individual photons and report the number of photons incident at a pixel (i.e., independent of the energies of the arriving X-rays). This capability is realized through additional processing hardware that analyzes the shape of the current pulse produced when X-rays interact with the detector material and determines the number of individual pulses produced. In this way, these detectors can be shot-noise limited if the X-ray energies of interest produce current pulses in excess of the other noise sources in the system, as one can simply set a threshold below which these smaller fluctuations can be ignored. Finally, energy-sensitive detectors record the number and energy of the incident X-rays at each pixel. This capability depends on fast electronics that analyze the detailed current pulse shape and height produced by each arriving X-ray; since X-rays of different energies produce current pulses with different amplitudes (which are sufficiently distinct from the signal produced by the arrival of multiple X-rays of identical total energy), one can therefore determine the X-ray energy up to an uncertainty limited by noise and choice of digitization.

As knowledge of the X-ray energy is critical in XRD, energy-sensitive detectors provide significant opportunities. For example, measuring the energy of the scattered X-rays allows one to avoid spectral filtering, which otherwise reduces the signal intensity at both the desired and unwanted energy bands. It also allows one to work with broadband signals, which increases the system's photon throughput and relaxes constraints on required X-ray optics and/or the spatial configuration of the detector. In order for energy sensitivity to be useful, however, the detectors must have a fractional energy resolution, $\Delta E/E$, that does not limit the momentum transfer resolution [see Eq. (4)]. However, the choice of energy resolution requires careful analysis of all of the underlying components, such as the properties of the active detector material itself and the associated processing electronics (including the application-specific integrated chip, or ASIC). While a smaller energy resolution is desirable from a resolution perspective, it is also critical to realize sufficient SNR in the available (usually short) exposure times. Unfortunately, the energy resolution and count rate of energy-sensitive detectors are inversely related: Doing a good job in characterizing the properties of each incident photon limits the rate at which those photons should arrive at the detector. Thus, there typically exists an optimal combination of energy resolution and count rate for a particular CA-XRDT configuration and application.[66]

Equally important to measuring the energy of the X-ray, one must measure accurately the scatter angle. To accomplish this, the detector pixel size should be small enough that one adequately samples the magnified, projected-coded aperture pattern. While Nyquist sampling is technically sufficient, better performance is typically realized with finer sampling, especially in the case of highly multiplexed signals. It is also important to note that larger pixels result in both a larger angular and energy uncertainty in the detected X-rays, which negatively impacts the achievable momentum transfer resolution according to Eq. (4). Despite the aforementioned advantages of smaller pixels, there is typically a tradeoff between pixel size and performance. For example, smaller pixels often result in lower SNR, since they collect less signal but have comparable (or sometimes larger) noise compared to larger pixels. On the other hand, energy-resolving detectors make use of the so-called small pixel effect[67] to realize superior energy resolution but, at the same time, the amount of charge sharing between pixels (i.e., when an X-ray partially deposits its energy in multiple adjacent pixels) can lead to artifacts and extra uncertainty in the measured spectra. While various hardware[68] and post-processing[69] corrections have been demonstrated, they typically do not completely solve the problem.

Beyond the actual size of the pixel, the spatial arrangement of detector pixels is also important. As hinted at in Eq. (5), maximizing the number of code features projected onto the detector generally increases the system resolution. Since coding predominantly occurs in the spatial domain, this generally argues for a spatially extended detector array (e.g., a 2D detector subtending the anticipated range of scatter angles). However, it can be sufficient to use sparse or random arrays of pixels due to the distributed nature of the scatter signal[70,71] (with the caveat that highly crystalline materials give textured scatter that is more concentrated at certain locations). In practice, detectors typically are arranged into linear or rectangular multi-pixel arrays. However, the robustness against bad pixels means that less perfect, and therefore less expensive, crystals can be sufficient for CA-XRDT. Thus, while an energy-sensitive area detector would be ideal for most CA-XRDT applications, the cost, performance specifications, and availability of the detectors must also be considered. While large, high-performing two-dimensional energy-sensitive detectors are beginning to emerge, current off-the-shelf technology typically requires one to face a tradeoff between using large area energy-integrating detectors or energy-sensitive detectors with limited spatial extent and/or packing density. The final choice of detector therefore depends on the specific application and associated design constraints.

1.3.1.4 Computational Infrastructure

Given that CA-XRDT is inherently a computational imaging modality, the final piece of any CA-XRDT system is the computational engine (i.e., both hardware and software) associated with converting the raw measurements into a description of the object and/or its underlying properties. While some limited system configurations allow simple Fourier processing of the data, a model-based approach to data processing is typically required. As discussed in the next section, this involves modeling the geometry and physics of the system at the global and component levels so that the object can be faithfully estimated from the complex, noisy, multiplexed measurements. To this end, the model must be as precise and accurate as possible, with residual model error potentially providing a practical limitation to overall system performance. The level of sophistication (e.g., included physical phenomena) and accuracy (e.g., choice of sampling, interpolation, etc.) of the model depend on the particular system of interest but some degree of experimental calibration, component characterization, and cross-validation with real measurements is required.

CA-XRDT data can be very high-dimensional. For example, objects of interest may exist in up to three spatial dimension, may vary in time, and have a momentum transfer (i.e., material) dimension as well. Similarly, detectors may record the time, location (in multiple dimensions), and energy of each of the scattered X-rays. Thus, while CA-XRDT gives one access to large quantities of data, real-time processing of data from large, finely sampled objects can be prohibitive both in terms of computational time and resources. While advances in CPU and GPU processing capacity continue to reduce this challenge, it is important to explicitly consider the computational requirements in the overall CA-XRDT system architecture. This co-design of hardware (e.g., source and detector properties, code pattern, etc.) and software (model accuracy, estimation algorithms, etc.) allows one to design and optimize the entire system to make optimal use of the available components.

1.3.2 Modeling and Estimation

1.3.2.1 Forward Model

A forward model is a description of how points in the object space are mapped into the measurement space. This mapping can, in general, be nonlinear in the object. While including such nonlinearities is important for realizing the most accurate simulation of a particular object, it does not generalize to predictions of measurements for arbitrary combinations of objects and, therefore, complicates the estimation process. We instead seek a linear (or linearized) model in which the overall scatter signal produced by a protean object is simply the sum of the signals from each point in the object.

For the configuration shown in Figure 1.7a, the number of scattered X-rays g with energy E reaching a detector at location r_d at time t is

$$g(E,\mathbf{r}_d,t) = \int \mathbf{d}\mathbf{r}_o\, \mathrm{d}q\, \Phi(E)\frac{\mathrm{d}\sigma(E,t)}{\mathrm{d}\Omega}\,\Delta\Omega_d\; T_s(E,\mathbf{r}_o,t)\; T_d(E,\mathbf{r}_o,\mathbf{r}_d,t)$$

$$\delta\left(E - \frac{hcq}{\sin\left(\dfrac{\theta}{2}\right)}\right), \tag{1.7}$$

where the solid angle subtended by the detector element is given by $\Delta\Omega_d$ and $\Phi(E)$ is the incident source spectrum. Note that the object, $f(\mathbf{r}_o,q)$, is implicitly contained within the differential scatter cross section, which may be viewed as time-varying if there is relative motion between the source and object or evolution of the object itself. Here T_s and T_d describe the transmission of X-rays from the source to a particular point in the object and from the object to a location at the detector, respectively. These transmission functions potentially include the absorption of X-rays both from within the object (i.e., self-attenuation) and as a result of passing through a coding or collimating optical element. Here we explicitly assume that this transmission can be time-varying, such as in the case of a moving object or dynamic code element. In order for the model to be linear, however, we typically assume that the optical depth of the object is small (i.e., self-attenuation is negligible) or constant although one can overcome this limitation through iterative application of the linear model. Finally, the delta function enforces Bragg' law and relates the scatter angle and momentum transfer value to the X-ray energy, which is valid for the case where texturing is negligible. The total signal arising from an object is the integration of the scatter from all object dimensions (both in position and momentum transfer).

Beyond the direct scattering process, one can also include a transduction process that converts the uncorrupted, noise-free, mean counts predicted in Eq. (7) into more realistic signals at the detector. Such a transduction process can include effects like finite DQE, uncertainty in the determination of an X-ray's energy (i.e., finite energy resolution), charge sharing between pixels, nonlinear conversion between incident and registered photon counts, detector and photon noise, etc. In general, these effects can be accounted for by applying the appropriate mathematical operations to the idealized model above.

While Eq. (7) is written as an analytic integral, the integral cannot, in general, be evaluated analytically. It is therefore necessary to discretize the model for both computer-based implementation and numerical analysis. We note, however, that one can alternatively generate a forward matrix either through other simulation methods (e.g., Monte Carlo) or via experimental determination (i.e., measuring the signal for a point object at each location) although these methods can be cumbersome, impractical, or impossible and lacking generalizability. Regardless of the method, determining the impulse response for every object within the desired system FOV enables one to generate a forward matrix H such that

$$g = Hf,$$

where g and f are vectors of length N and M, respectively, and H is an $M \times N$ matrix. Depending on the measurement architecture and code structure, one can operate in a regime where the number of object points is less than, comparable to, or greater than the number of distinct measurements.

One can use the forward matrix for several purposes. First, one can use the model to perform design studies. For example, one can determine an estimate of the system resolution through evaluation of the Gram matrix or the robustness of the system to preserve information in the presence of noise through conducting a singular value decomposition (SVD) analysis.[59] Alternatively, one can perform analyses of image and/or detection performance over an ensemble of generated objects.[72,73] Finally, one can simply calculate the expected signal for a particular object in order understand the shape and amplitude of the expected signal. This enables cross-validation between simulated and empirical data and acts as a critical calibration step for minimizing model error between virtual and as-built systems.

1.3.2.2 Estimation Algorithms

The process of estimation (also known as reconstruction or inversion) involves determining f from a measurement of g. In the absence of noise and for the case of a well-conditioned system (e.g., an isomorphic mapping between object and measurement space), one can simply calculate the inverse H^{-1} and apply it directly to the measurements. However, simple inversion in the presence of noise can lead to an amplification of the singular values that carry the least information and are most likely to be corrupted by noise. In order to provide robustness against this, one can filter/denoise the directly data[74] or choose a regularized reconstruction that preferentially "ignores" aspects of the signal that are most likely due to noise.[75] One can also use regularization and/or prior information for estimation in the case when the system is poorly conditioned,[76] such as when there exists more than one possible object that is consistent with the measurement. This "extra" information provided to the estimation algorithm effectively nudges the solution in a particular direction so that one ends up with the correct solution out of the multitude of possible solutions.

A general class of such problems arise in compressed sensing (CS) scenarios, in which $N<M$ (i.e., the object dimensionality exceeds the measurement dimensionality). While this leads to an underdetermined estimation problem, it has been shown that the use of prior information, such as promoting sparsity or smoothness in a particular basis, can result in high-fidelity estimates of the object.[77] Examples of CS algorithms that have been used

or developed for CA-XRDT include TWIST,[78] TV,[76] and group TV.[79] In some cases, one can actually employ the data itself to choose the preferred regularization parameter,[80] rather than tuning it by hand for optimal performance. Through successful use of such CS algorithms, one can extend the space of objects that coding applies to from sparse, bright point objects (as in the case of X-ray astronomy or discrete crystals in XRD) to extended, heterogeneous, natural objects (as in the case of security or medical imaging).

As discussed above with respect to computational resources, the estimation process should also be included in the overall system design. For example, Odinaka et al.[81] showed that estimation algorithms can be greatly accelerated if one chooses to employ a measurement system with sufficient symmetries. Furthermore, certain system configurations are more amenable to the use of CS-type algorithms, which can reduce measurement resources (e.g., detector costs, scan time, or dose[59]) without compromising image quality. In addition, one can design complimentary measurements that provide side-information (e.g., a measurement of the object's attenuation properties) that improve estimation fidelity by improving the system conditioning and/or constraining the estimation through acting as a relevant prior.[82,83] Finally, the structure of the algorithm itself can be implemented in such a way that it can be accelerated through parallelization on one or more GPUs. In this way, the throughput gains achieved in the physical sensing process can be maintained during the post-processing steps.

1.3.2.3 Classification Algorithms

For detection problems, one may desire to go beyond estimation and use the recovered $f(r_o, q)$ distribution to identify or classify the material at each location in the object. The traditional method for performing material identification involves comparing each recovered form factor to a library of previously acquired form factors whose associated materials are well-known. While the metric of comparison may take a variety of forms, such as Euclidean distance, cross-correlation, Earth Mover's Distance, etc., the goal is to find the material in one's library that is most similar to the estimated object material. If the similarity is strong enough, then one can claim successful identification of the material, and the auxiliary properties of the material (e.g., whether it is a threat or non-threat material in a security application) also become known.

In some cases, however, the form factor may not match well with any of the elements in the library. For example, if the spatial resolution of the system is such that multiple materials are present in a single voxel, then the resulting form factor is the weighted linear sum of the materials present. In this case, one can apply spectral unmixing[84] techniques to determine both the fractions and materials present. A more challenging problem, however, arises when the material present in the object simply does not appear in the library. Given the multitude of materials that one might find in certain applications and the corresponding dependence of the form factor on environmental factors (e.g., temperature, strain, processing history, degree of crystallinity, etc.), this scenario may arise in various applications. To mitigate this situation in the case of detection (i.e., rather than material identification or quantitative material analysis), Zhao et al.[85] have shown how the machine learning algorithms can be used to identify key features of the form factors within a given class of interest so that new instantiations of known materials or altogether new materials can still be properly classified.

1.4 Snapshot CA-XRDT Techniques

One particularly useful advantage of CA-XRDT techniques are that they enable snapshot imaging throughout a volume. Since only a single, static measurement is required, the measurement can be fast and requires no moving parts. The dimensionality of the estimated volume depends on the illumination and detection volumes, sensor properties, and object extent. In this architecture, optics are placed between the source and object to define the region of the object to be illuminated, and a detector-side coded aperture is placed between the object and detector. In this section, we discuss several snapshot architectures that have been demonstrated for a variety of object types.

1.4.1 Pencil Beam

Snapshot pencil beam XRDT is a particularly interesting imaging modality. The idea is that one illuminates an object with a single pencil beam and is able to recover the XRD form factor at each location along the beam (see Figure 1.10). This allows for spot tomography, where only a particular region of interest is illuminated (in contrast to CSCT, for example, which requires illuminating an entire plane at minimum). As a result, the required dose is significantly reduced due to the elimination of unnecessary dose to neighboring regions, and it has been shown that CA-XRDT performs better at fixed dose than DT.[59] Furthermore, pencil beam XRDT has the advantage that each individual line through the object can be recorded and reconstructed independent from one another, thereby reducing the computational requirements.

The first experimental demonstration of CA-XRDT was performed using an incident pencil beam with a narrow, filtered spectrum.[8] A two-dimensional energy-integrating detector (CsI flat panel) was used to collect the scatter signal modulated by a one-dimensional, binarized sinusoidal mask. With this choice of components, the system therefore operated in an angle-dispersive mode and enabled estimation of the XRD form factor at each point along a single line through the object. While spectral filtering reduced the scattered X-ray flux, the use of a large area detector and a code with 50% transmission

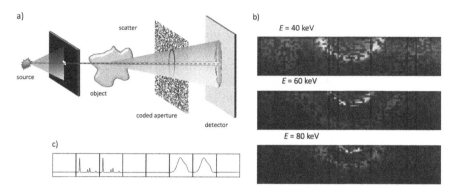

FIGURE 1.10
(a) Schematic of a pencil beam CA-XRDT setup. A pinhole collimator selects a particular, limited bundle of rays that intersect the object. The resulting scatter is modulated by coded aperture and measured on a detector. (b) Experimental scatter data from a thin sample of Teflon measured in a pencil beam CA-XRDT system with a 2D, energy-sensitive detector. The lower half of the Debye rings are clearly visible, along with a crisp modulation from the mask. (c) Representative illustration of the recoverable object information: a form factor at each location along the pencil beam direction, where each block represents a voxel in the object.

allowed MacCabe et al. to capture almost half of the generated scatter. Furthermore, the coded aperture uniquely modulated each location along the pencil beam, thus relegating any need for serial scanning. The resolution for the as-demonstrated system was approximately 3 cm along the axial (i.e., beam propagation) direction and 0.07 nm^{-1} in momentum transfer (corresponding to about 3% fractional uncertainty), with exact values depending on the extent of the object.[86] It is worth noting, however, that this was simply a proof of concept experiment and that significantly improved imaging performance can be achieved with alternative coding strategies and implementations. Furthermore, as the focus of many of these experiments are in the security space (i.e., detecting explosives in baggage), the systems were designed and built to run in the hard X-ray regime (i.e., up to 160–200 keV). As a result, the smallest code feature size is limited to 1–2 mm, which ultimately limits spatial resolution.

As a follow-on experiment, Greenberg et al. showed that the one could achieve the same snapshot pencil beam imaging capabilities by using instead a linear array of energy-sensitive detector pixels. Thus, the two-dimensional object is mapped onto the two-dimensional detector (i.e., one energy and one spatial dimension now), and the architecture can be interpreted as operating simultaneously in ADXRD and EDXRD modes. Note, however, that coding is typically only applied to the spatial dimension but can consist of one- or two-dimensional codes. This configuration requires no spectral filtering and, therefore, makes maximal use of the generated X-rays within the pencil beam. In addition, it is spatially compact (relative to the need for a two-dimensional detector) and inherently low noise, since energy-sensitive detectors are typically shot-noise limited. The axial image resolution was demonstrated to be on the order of 5 mm and the fractional momentum transfer resolution was generally less than 10%, which was limited by the 10% fractional energy resolution of the detectors.[87] While the axial resolution is comparable to DT, it is slightly larger than conventional CSCT approaches. However, the overall system resolution is sufficient to enable one to determine subtle material properties, such as the fractional composition of liquid mixtures and to perform material classification of items placed along the beam relative to a known library.[9]

One can use the CA-XRDT approach to further reduce the hardware resources required to record $f(q,z)$ by choosing to operate in a compressive sensing mode.[88] Rather than matching the object and detector dimensionalities, one can alternatively use only a single energy-sensitive detector element with sub-pixel coding. In this configuration, the task is to recover the two-dimensional object using a one-dimensional measurement (i.e., just a measurement of the scattered X-ray energy spectrum). While this represents a degree of compression in embedding space, one can also seek to simply recover more object features than the number of distinct measurements. For objects that are properly sparse in an appropriate basis, it was found that this single pixel scheme is able to determine accurately the location and form factor of an unknown object up to a compression ratio (i.e., M/N) of approximately 20.

1.4.2 Fan Beam

While spot tomography with a pencil beam can be advantageous in some situations, imaging larger object volumes using raster scanning of the pencil beam can be slow and cumbersome. Therefore, fan beam CA-XRDT strategies have been developed to image $f(q)$ throughout a planar slice of an object via a single measurement.[56,89] In this case, one places a slit collimator (or focusing optics, if preferred) between the source and object to produce a fan beam. Each ray in the fan beam produces a set of rings of diffracted X-rays (as in the case of the pencil beam described above), which overlap with one another in angle and energy. The scatter from different rays can be distinguished both by virtue of their unique coding (since it passes through slightly different regions of the coded aperture) and the fact that the scatter rings from each ray are centered on different locations (due to the relative shift and/or tilt of the different illuminating rays). A general example of fan beam CA-XRDT is shown in Figure 1.11.

Using this illumination strategy, multiple choices of coding and detector configurations have been demonstrated. For example, MacCabe et al.[89] used spectral filtering combined with a two-dimensional code and energy-integrating detector to demonstrate an ADXRD version of fan beam CA-XRDT. Alternatively, Hassan et al.[56] showed that a linear array of energy-sensitive detectors with a one-dimensional code can operate in an ADXRD/ EDXRD mode and is sufficient for compressive recovery of reasonably complex objects. However, the robustness of the method to object complexity and

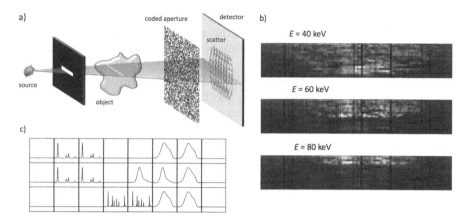

FIGURE 1.11
(a) Schematic of a fan beam CA-XRDT setup. A slit collimator selects a particular set of rays that intersect the object. The resulting scatter is modulated by coded aperture and measured on a detector. (b) Experimental scatter data from a high density polyethylene (HDPE) rod measured in a fan beam CA-XRDT system with a 2D, energy-sensitive detector. The extent of the object can be seen in the convolution of the rings along the horizontal direction, and it can be observed that the modulation contrast of the coded aperture is less obviously visible (although still present). (c) Representative illustration of the recoverable object information: a form factor at each location in the plane of the fan beam, where each block represents a voxel in the object.

clutter improves if a two-dimensional code and two-dimensional, energy-sensitive detector are used.[24,90] Compared to pencil beam CA-XRDT, the fan beam configuration has comparable imaging performance and a reduction in scan time commensurate with the extent of the fan beam (e.g., a 100 mm wide fan beam presents a 100x speedup relative to a raster scan of a 1 mm pencil beam across the same region).[56] This approach is therefore particularly useful for imaging the entirety of a quasi-planar object or a targeted slice through a thick object. One can, of course, always image larger volumes by raster scanning the object using linear translational motion in a direction normal to the fan beam plane.

1.4.3 Cone Beam

The family of snapshot CA-XRDT configurations can, perhaps unsurprisingly, be extended to a cone beam mode in which the full four-dimensional object (3 spatial + 1 momentum transfer dimensions) is imaged in a single measurement. To realize cone beam CA-XRDT, a two-dimensional code with both shift and scale variance is required. Similarly, while one can choose to use either a spectrally filtered source or energy-sensitive detection, one must use a two-dimensional pixelated detector. It is worth noting that this scheme is intrinsically compressive in the object embedding dimension, as the object dimensionality necessarily exceeds the detector dimensionality. As in the case of the fan beam configuration described above, though, cone beam CA-XRDT works because each ray passing through the object produces a uniquely positioned and modulated scatter signal at the detector. However, an additional restriction arises in the design of such a system: the angular extent of the usable cone beam is limited by the fact that the bright primary beam must not significantly overlap with scatter signal. In practice, this requires that the divergence angle of the cone beam be less than the smallest desired scatter angle. Furthermore, the absolute locations of the object and detector must be such that the scatter from the center of the cone has a sufficient propagation distance to escape the footprint of the primary beam. It is generally advantageous, then, to place the object fairly close to the source and to work at lower energies to maximize the deflection angle of the scatter. For many configurations, this enables one to image a conical volume with a cross section of up to several centimeters over a range of tens of centimeters. Cone beam CA-XRDT is therefore best suited for imaging smaller objects with the advantage that, if the object fits within the cone beam, no raster scanning is needed.

Figure 1.12a illustrates the configuration for cone beam CA-XRDT, in which a limited cone of X-rays fully illuminates an object. The superposition

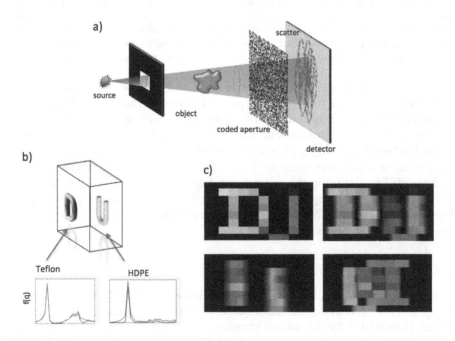

FIGURE 1.12
(a) Schematic of a cone beam CA-XRDT setup. An extended area collimator selects a suitable range of rays that intersect the object. The resulting scatter is modulated by coded aperture and measured on a detector. (b) Schematic of a simulated phantom consisting of the letters "D" and "U" composed of Teflon and HDPE, respectively. (c) Reconstructions based on simulated data, showing different perspectives of the phantom described in (b). The colors silver and copper correspond to Teflon and HDPE, respectively, as determined by classifying the estimated form factor at each location against a library of known materials.

of scattered X-rays, generated from locations throughout the object, is modulated by the coded aperture and arrives at an area detector. To demonstrate the feasibility of this system, Greenberg et al.[24] created a simulated phantom consisting of an object whose material properties vary throughout the illuminated region, as shown in Figure 1.12b. For this configuration, they used a numerical model to simulate the raw scatter signal, used the data and forward model to estimate the object, and applied a correlation-based classifier to identify the material at each location in the object volume. The resulting four-dimensional object is displayed in Figure 1.12c, where different perspectives of the phantom are shown, using color to indicate the material present at each location. The success of this numerical study indicates the potential for this method.

1.5 Multi-Shot CA-XRDT Techniques

The snapshot CA-XRDT techniques described above offer the simplicity of requiring only a single measurement and the advantage that one can select only specific regions or volumes of interest to image; however, requiring snapshot operations results in several limitations. For example, a single measurement can record only a limited amount of information. As a result, one must often choose between maximizing throughput (e.g., by increasing the illuminated volume of the object) or realizing good system conditioning. Because most configurations are biased toward the latter, many of the generated source X-rays are traditionally unused, and extending the imaged FOV requires multiple, sequential measurements (e.g., raster scanning). In addition, performing tomography from a single perspective results in fundamental resolution limitations and potential challenges associated with dynamic range, such as in the case of objects with widely varying scatter strengths. To improve both image quality and overall X-ray photon throughput (which, correspondingly, reduces scan time and/or associated dose), we consider two examples of multi-shot CA-XRDT methods.

1.5.1 Multi-View CA-XRDT

One method for extending the previously discussed snapshot CA-XRDT approaches involves acquiring multiple such measurements with rotations of the object between successive acquisitions. The hardware and configuration is therefore nearly identical to the snapshot techniques described above: a source-side optic defines the illumination volume, a detector-side coded aperture modulates the scattered signal, and the object (or source, detector, and coded aperture) is placed on a rotation stage. Multiple, snapshot-like measurements are acquired for the object in different orientations, and estimation is performed using the combined, multi-view data set (see Figure 1.13). While each measurement provides a spatially resolved estimate of the XRD form factor, additional measurements provide an increasingly tighter localization of the scatter origin (especially in the direction of beam propagation) with an associated improvement in the momentum transfer resolution.

A variety of implementation configurations are possible, including pencil, fan, and cone beam illumination (with the appropriate shifts and rotations necessary to adequately cover the region of interest) in conjunction with the appropriate detection and/or spectral filtering. Holmgren et al. demonstrated multi-view CA-XRDT in a fan beam configuration.[60,91] For their experimental configuration, they observed spatial resolutions on the order of 1–2 mm in both the transverse and axial dimensions, with a 10% fractional momentum transfer resolution (limited by the detector's energy resolution). More generally, they found that the imaging performance improves with increasing numbers of views if the total incident flux is allowed to also increase. In the case of a fixed total incident flux distributed across all views, however, they found that the spatial and momentum transfer resolutions improve as the number of views increase in the high SNR regime, but that there are diminishing returns beyond 5–10 measurements. The performance gains in going from one to two orthogonal views are most significant in this scenario, due largely to the asymmetric transverse and axial spatial resolutions of the single-view system. In contrast, it is actually preferable to use only a single measurement and avoid shot noise-related uncertainty in the low SNR regime.

The presence of a coded aperture significantly improves the overall system performance for all SNR and view numbers as compared to the uncoded case. Thus, in the interpretation of this measurement approach as a form of limited-view, compressive CSCT (rather than multi-view CA-XRDT), it can

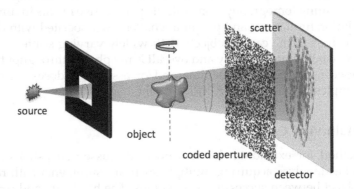

FIGURE 1.13
Schematic of a multi-shot CA-XRDT setup. As in snapshot CA-XRDT, a single coded aperture between the object and detector modulates the signal; however, one now includes relative motion between the object and detector and measures the signal repeatedly as a function of time while the system evolves. As an example, we include here the case of cone beam CA-XRDT with rotation about a transverse axis.

be seen that the improved conditioning provided by the code outweighs the fact that approximately half of the X-rays are blocked by the code, which is consistent with the results of Kaganovsky et al. for coded, compressive CT.[92] This opens up potential applications for realizing simultaneously high-throughput and high-resolution XRD-based imaging.

1.5.2 Structured Illumination XRDT

An alternative method for exploiting the temporal measurement degree of freedom in a CA-XRDT system involves placing only a coded aperture between the source and object.[65] This effectively modulates, or structures, the illumination such that an XRD signal is produced at all locations in which the illumination and object volumes overlap. This configuration allows one to record simultaneously the attenuation signal from all of the primary incident rays, and the scatter can be measured at all other, interstitial locations. By moving the object through the structured beam along a known trajectory and recording the time-dependent scatter signal, the entire volume of interest can be imaged. One can choose linear translational motion (as in the case of a conveyor belt), rotational motion (as in a CT configuration), or a combination thereof. As opposed to simple raster scanning, each voxel in the structured illumination scheme is illuminated multiple times and its scatter is multiplexed with that of nearly every other voxel. In this way, the source-side coded aperture effectively patterns the scatter signal in time (rather than in space or energy, as in the previous configurations discussed). To illustrate how this coding allows one to demultiplex the resulting signal, consider an illumination pattern diverging from a single focal location. The relative frequency of the time-dependent scatter signal from a given voxel depends on its axial location (along the source-to-detector direction), and the specific pattern of the modulation depends on the transverse location of the object. Together, this produces a unique spatio-spectral-temporal signal at the detector for each object voxel. Consequently, the spatial and momentum transfer resolutions are now governed largely by the temporal and energy resolution of the measurement system, respectively.

An example of a generalized structured illumination XRDT system is shown in Figure 1.14. In this scheme, the cone beam emitted from single source is modulated by a coded aperture located prior to the object. The motion of the object produces a scatter signal that multiplexes different regions of the object with one another in different combinations, such that the signal can be demultiplexed without the need for a detector-side code.

The main advantage of the structured illumination scheme is a dramatic increase in photon throughput. One can now use a large fraction (up to nearly half) of the X-rays emitted by a traditional, cone beam source, which produces a 10–100x increase in available signal relative to alternative collimation schemes and allows for joint ADXRD and EDXRD operation. Furthermore, the removal of the detector-side code effectively doubles the measureable

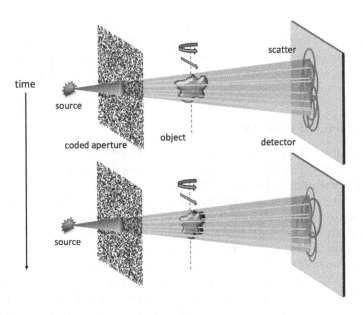

FIGURE 1.14
Schematic of a structured illumination CA-XRDT system. A source-side coded aperture is placed between the source and object, which creates a multitude of rays that illuminate the object. Relative motion between the object and source, such as rotation or translation, allows for the entire object to be illuminated. Furthermore, each voxel is, in general, illuminated multiple times and from different perspectives. The multitudes of scatter signals from each source-object configuration are collected at the detector, without the need for a detector-side code.

signal at the detector. Also, in contrast to conventional raster scanning, each voxel is illuminated multiple times and from different perspectives, resulting in a larger available signal for a fixed total number of incident X-rays and a better conditioned system, respectively. A second advantage is that one can use a much smaller detector area, as compared to schemes requiring a detector-side coded aperture; since the coding is now in time, the number of temporal channels replaces the need for many pixels distributed in space. Since the detector cost and complexity are often critical design parameters in an XRDT system, reducing the requirements on this key component makes the overall system simpler and, potentially, significantly less expensive. A final advantage is that one can easily vary the illumination structure and detector configuration to optimize the degree of compression (and degree of multiplexing) for a particular imaging or detection task. Relatedly, the scan times required for imaging a full three-dimensional volume and a two-dimensional planar slice through an object are nearly identical, subject to the constraint that the signal be sufficiently well-conditioned for inversion.

A proof of concept structured illumination XRDT experiment was conducted by Greenberg et al.[65] In this work, the authors illuminate a planar slice of an object with a structured fan beam (i.e., a fan beam passing through

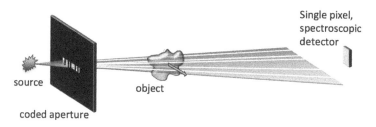

FIGURE 1.15
Schematic for the fan beam structured illumination CA-XRDT experiment described in Ref. 65.
An object undergoes linear translational motion through a modulated fan beam, and the scatter is collected by a single, spectroscopic detector pixel with temporal resolution. This configuration is sufficient to image the XRD form factor throughout the planar slice of the object that
is illuminated.

a one-dimensional coded aperture, see Figure 1.15). The object undergoes
linear translational motion in the plane parallel to the fan beam, which
simultaneously acts to illuminate all object voxels of interest and to provide
view diversity through the effective rotation of the object (relative to the fixed
focal spot of the source). The scatter is recorded by a single, energy-sensitive
pixel with temporal resolution. From this spectral-temporal measurement,
they recover the XRD form factor at each voxel in the two-dimensional slice
of the object. The axial and transverse spatial resolution were on the order of
10 mm and 1 mm, respectively, and the fractional momentum transfer resolution was less than 10% (limited, in this case, by the angular blur). Thus, the
imaging performance was comparable to that observed using snapshot fan
beam CA-XRDT; however, the required source flux (on a per unit detector
area basis) was over 10 times smaller, thus enabling significantly faster scan
times. In a follow-on study, Pang et al.[93] demonstrated full four-dimensional
imaging (i.e., three spatial and one momentum transfer dimension) through
the use of cone beam illumination with a two-dimensional source and
detector-side code.

While offering significant benefits, there are several disadvantages to the
structured illumination XRDT approach. For example, it is no longer possible to do spot illumination—the requirement of object motion to "fill in"
the non-illuminated regions implies a minimum of imaging a quasi-two-
dimensional region. Relatedly, the massive inherent parallelism of the measurement approach that leads to the speedups in data acquisition also leads to
non-trivial computational challenges, as the entire illuminated object volume
must be demultiplexed together. The maximum useful imaging volume is
therefore limited by the available memory and/or processing speed of the
associated computational engine associated with the system. Finally, with
high degrees of multiplexing, it is critical that the background noise in the
system be small. Because the bright primary beam is interacting with the
highly attenuating (and, therefore, highly scattering) source-side coded aperture, the potential for creating unwanted scatter signal is much greater than

that of a detector-side code (where only the dim scattered beam interacts with the coded aperture). Thus, it is critical to carefully implement the source-side coding.

1.6 Applications of CA-XRDT

X-ray diffraction has been used in a variety of scientific and commercial applications, including the characterization and analysis of materials in the medical, security, and industrial spaces. However, most previously employed XRD-based systems are used in a non-imaging mode that is sensitive primarily to surface structure (i.e., the topmost 50–100 μm in a standard diffractometer). While XRDT has long demonstrated promising new capabilities, efforts to employ it in practical applications have, historically, been unsuccessful due to practical limitations. The suite of CA-XRDT techniques discussed above overcomes these limitations and allows one to:

1. speed up XRDT and/or make use of simpler, less bright sources;
2. reduce the required dose by increasing X-ray detection efficiency and/or enabling targeted imaging;
3. relegate the need for rotation or other relative motion in situations where it is not possible; and
4. exercise a large degree of customization in the configuration to optimize the system performance to a particular task and/or constraint.

These new capabilities enable operational regimes that were previously not possible and makes practical the use of XRDT in scenarios in which it was previously precluded. While not a complete list, the remainder of the chapter is devoted to the discussion of specific areas of application in which CA-XRDT has already made or could make significant impacts relative to the previous state-of-the-art methods.

1.6.1 Security

While initially limited to metal detectors, X-rays have been used in security imaging for several decades now to go beyond simply detecting the presence of metal objects in identifying threats of various shapes and compositions.[94,95] The initial application of X-rays in the security domain focused mainly on identifying shape threats, such as guns or knives. More recently, significant efforts have been made to provide adequate detection of material threats and/or contraband, such as explosives or drugs, in which the detection class associated with an object is independent of its shape and completely depends on what material(s) it is composed of. X-ray diffraction is critical to this endeavor because, unlike transmission-based imaging, it provides a molecular signature of the object that can be uniquely associated with a particular material. To highlight the capabilities of XRD-based discrimination over conventional multi-energy transmission (e.g., CT) imaging, Figure 1.16 shows the locations of different materials in a two-dimensional, representative transmission and XRD space (assuming a 10% variation in the respective material properties for each material). The larger separation and lack of overlap between common materials in the XRD space indicates its superior capability for distinguishing materials from one another.[96] Beyond simply making use of XRD, though, the fact that such target objects are concealed in a cluttered environment means that XRDT specifically is required in order to isolate particular regions of interest within a bag.

XRDT is scientifically well-suited to the challenge of detecting concealed materials of interest because of its sensitivity to the molecular structure of a material; however, the practical and logistical constraints on such systems have made their development and adoption challenging. In particular, a checkpoint XRDT system must have a small footprint, low cost, low power consumption, and run in real-time (several seconds per bag, including data acquisition and processing). Several commercially developed XRDT systems have been developed and deployed to date but none have met all of the requirements of such a system. For example, the checked bag HiScan 10065 HDX[34,97] and XRD 3500[32] systems both employ an initial, upstream CT scanner to identify potential threats and a downstream, targeted (i.e., scanned pencil beam) DT XRDT system to help resolve the alarms. Because of the required CT front-end and the choice of using a single-beam DT XRDT

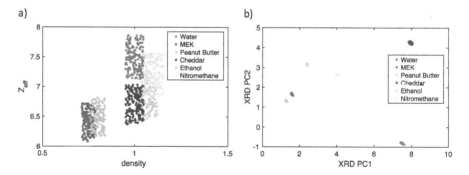

FIGURE 1.16

Two-dimensional representation of six materials commonly found in luggage in (a) transmission space (shown here in terms of density and effective atomic number, Z_{eff}) and (b) XRD space (shown here in terms of the first and second principle components). X-ray diffraction space allows for better, easier separation between the materials. We have assumed that each material has a 10% uncertainty in each of its effective dimensions (for XRD, the uncertainty was applied to each momentum transfer value in the relative form factors, prior to principle component analysis).

configuration, the cost, size, and power requirements for these systems are not appropriate for operation at the checkpoint. The XDi system, in contrast, uses XRDT as a primary means for evaluating every voxel of each bag and conforms to the required checkpoint footprint.[33] The XDi machine has demonstrated low false alarm rates for solid explosives and multiple liquid explosive threats in containers and inside bags, allowing both liquids and heavy electronics to remain inside carry-on baggage. However, the use of an inverse fan DT configuration requires an expensive, complex multi-focal spot X-ray source and low bag throughput.

As discussed above, the CA-XRDT approach's simple configuration and high photon throughput can provide a path toward the practical implementation of XRDT at the checkpoint. To this end, Smiths Detection and Duke University worked together to develop a prototype scanner based on snapshot fan beam CA-XRDT,[98] as shown in Figure 1.17. Using this configuration, the XRD form factor at each voxel within a planar slice of the bag is imaged with each measurement. As the bag moves along the belt, the entire volume of the bag can be scanned in a slice-by-slice manner. This system has demonstrated the potential for improving detection and false alarm rates, especially for bulk threats. The additional information provided by XRD (as shown in Figure 1.17d), as compared to that available via multi-energy transmission imaging (see Figure 1.14c), has been shown to be particularly useful in accurately assessing the threat status of organic materials, liquids, and other materials with potentially large ranges of density (due to processing or packing). Perhaps just as important, however, is the fact that the use of the coded aperture approach enables the system to be built entirely with conventional, off-the shelf components, because it fits within the frame of a currently deployed AT machine and requires only wall power to scan bags in near real-time. Furthermore, the CA-XRDT configuration removes the need for moving parts (other than the belt) and naturally allows for imaging of a large FOV (both in terms of space and momentum transfer dimensions) with high photon throughput and limited detector extent (as compared to DT).

At a smaller scale, CA-XRDT is also well-suited for isolated, composite items. For example, one can perform bottle liquid scanning (BLS) to determine the material present inside of a container, which is currently of interest with respect to a range of venues, including airports, sporting events, and secure facilities. In contrast to current BLS systems that use Raman scatter or dual energy transmission X-ray imaging, an XRDT-based system can identify the contents of thick, opaque containers in a manner that is independent of the composition of the container itself.[24]

1.6.2 Medical Imaging

The discovery of X-rays was concomitant with their application to medical imaging over 100 years ago. Despite the development of CT and multi-energy detection, the reliance of medical imaging on photoelectric absorption

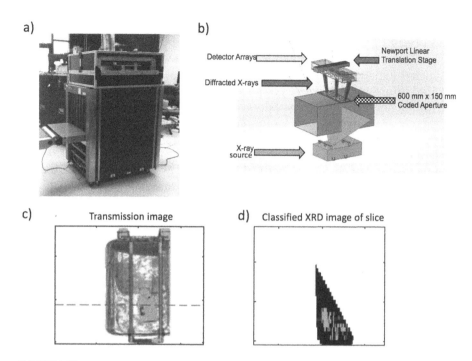

FIGURE 1.17
(a) Photograph of the prototype fan beam CA-XRDT system built in collaboration between Duke University and Detection, which is based on the frame of a 6046 scanner. (b) Schematic representation of the system, showing the X-ray source collimated to a fan beam, the tunnel in which the bag resides and the coded aperture (mounted at the top of the tunnel). The detectors have a finite standoff from the top of the tunnel and consist of a linear array of energy-sensitive pixels mounted on a translation stage. The system can produce (c) a multi-energy transmission image of the bag and (d) a map of the threat status (i.e., dangerous, red, or non-dangerous, green) of the materials located within a planar slice of the object (as indicated by the red dashed line in (c)).

necessarily limits the contrast and sensitivity that can exist in such images.[99] Different types of soft tissue or chemical compounds can have very similar densities and/or effective atomic numbers, thus fundamentally limiting the capabilities of transmission imaging in the absence of exogenous contrast agents. The structural information revealed via XRD has been shown to reveal important and otherwise unavailable information in a variety of situations. As discussed in more detail in Chapter 4 of this book, for example, XRD has been used to perform densitometry and mineral analysis of bone, marrow, and bone lesions,[100] conduct compositional analysis for disease detection in urinary and kidney stones,[101] blood,[102] and brain tissue,[103] study micro-calcifications as a precursor and indicator for disease in breast cancer,[104] and distinguish tissue types in cancerous and benign tumors[105–108] for margin assessment and/or in-vivo diagnosis. In general, these studies show that biological changes result in significant structural modifications of the tissue that are readily apparent via XRD but have traditionally been limited to non-imaging investigations of the scatter properties of a given sample; any application to in-vivo diagnostics of sub-surface targets or to situations in which the spatial distribution of a particular molecular structure provides required information, however, requires imaging of the XRD form factor (i.e., measurements at different, spatially resolved locations).

Coherent scatter computed tomography, DT, and CA-XRDT have all been used previously to study biological tissue although most studies have been limited to evaluating phantom objects. Coherent scatter CT has traditionally been the method of choice for medical imaging of ex-vivo samples because of the good achievable spatial resolution. Scans typically take long times (often on the order of hours), however, and the method is avoided for in-vivo application because of the high and unnecessary dose outside the target region of interest. Direct tomography has been developed for targeted analysis of samples but is also too slow for the investigation of larger volumes. The advantage of CA-XRDT in medical imaging is that one can perform efficient imaging of a region of interest ranging from a single line to a full volume. For ex-vivo applications, this can result in fast scans of large sample areas as is required, for example, for use in pathology.[87] More importantly, the ability to illuminate only the region of interest and collect a large fraction of the scattered X-rays make CA-XRDT a prime candidate for in-vivo imaging, where minimizing dose while maintaining resolution and sensitivity are of primary concern.

1.6.3 Industrial and Environmental Inspection

Another area in which XRDT has great potential is in the evaluation of commercial materials and products. In this space, the critical design parameters for such an XRDT system are typically the speed at which it can work (to ensure high product throughput), ease of integration into the existing framework, and overall system cost. CA-XRDT is compatible with these requirements because of its compactness, flexibility and customizability, and high-throughput capabilities. Although not yet deployed commercially, CA-XRDT has been explored for use in pharmaceutical inspection[109,110] of heterogeneous samples within a wide range of container types. Similarly, it can be applied to the food processing industry to perform real-time monitoring of fat content or detect organic contaminants inside food products (e.g., plastic shards inside of a chicken breast). Relatedly, the ability to identify different materials (and chemical species thereof), enables XRDT to be used for environmental monitoring. For example, Greenberg et al. used CA-XRDT to analyze the spatial distribution of copper nanoparticles at the parts per million level in wood samples to analyze the degree of leaching of different copper species from the wood into the environment.[111] In contrast to conventional methods that involve dissolving or otherwise modifying the samples of interest, the CA-XRDT technique is non-destructive and provides spatial (rather than bulk) concentration information. The application of CA-XRDT to such problems is still nascent but appears very promising due to its ability to provide spatial information (rather than simply bulk chemical composition) and can be customized and optimized to a particular task.

1.7 Conclusions

X-ray diffraction tomography is a powerful method for performing multi-dimensional imaging of the composition throughout the interior of an object. While XRDT has been around since the 1980s in various forms, the use of physical layer coding combined with model-based (and potentially compressive) data processing algorithms enable the new class of CA-XRDT systems. This approach allows for orders of magnitude increases in system throughput while maintaining the resolution of the system and can be realized using off-the-shelf, existing components. This combination of enhanced performance and simplicity has made CA-XRDT an exciting candidate for application to security scanning, medical imaging, and industrial inspection.

References

1. Harding, G., Kosanetzky, J. & Neitzel, U. X-ray diffraction computed tomography. *Medical Physics* **14**, 515–525, doi:10.1118/1.596063 (1987).
2. Harding, G., Newton, M. & Kosanetzky, J. Energy-dispersive X-ray diffraction tomography. *Physics in Medicine & Biology* **35**, 33 (1990).
3. Hounsfield, G. N. Computerized transverse axial scanning (tomography): Part 1. Description of system. *The British Journal of Radiology* **46**, 1016–1022, doi:10.1259/0007-1285-46-552-1016 (1973).
4. Benacerraf Beryl, R. et al. Three- and 4-dimensional ultrasound in obstetrics and gynecology. *Journal of Ultrasound in Medicine* **24**, 1587–1597, doi:10.7863/jum.2005.24.12.1587 (2005).
5. Lauterbur, P. C. Image formation by induced local interactions: Examples employing nuclear magnetic resonance. *Nature* **242**, 190, doi:10.1038/242190a0 (1973).
6. Lazzari, O., Jacques, S., Sochi, T. & Barnes, P. Reconstructive colour X-ray diffraction imaging - a novel TEDDI imaging method. *Analyst* **134**, 1802–1807, doi:10.1039/B901726G (2009).
7. Bunaciu, A. A., Udriştioiu, E. g. & Aboul-Enein, H. Y. X-ray diffraction: Instrumentation and applications. *Critical Reviews in Analytical Chemistry* **45**, 289–299, doi:10.1080/10408347.2014.949616 (2015).
8. MacCabe, K. et al. Pencil beam coded aperture X-ray scatter imaging. *Optics Express* **20**, 16310–16320, doi:10.1364/OE.20.016310 (2012).

9. Greenberg, J. A., Krishnamurthy, K. & Brady, D. Snapshot molecular imaging using coded energy-sensitive detection. *Optics Express* **21**, 25480–25491, doi:10.1364/OE.21.025480 (2013).
10. Brady, D. J., Mrozack, A., MacCabe, K. & Llull, P. Compressive tomography. *Advances in Optics and Photonics* **7**, 756–813, doi:10.1364/AOP.7.000756 (2015).
11. McCollough, C. H., Leng, S., Yu, L. & Fletcher, J. G. Dual- and multi-energy CT: Principles, technical approaches, and clinical applications. *Radiology* **276**, 637–653, doi:10.1148/radiol.2015142631 (2015).
12. Agatston, A. S. et al. Quantification of coronary artery calcium using ultrafast computed tomography. *Journal of the American College of Cardiology* **15**, 827–832, doi:10.1016/0735-1097(90)90282-T (1990).
13. McCollough, C. H. et al. Strategies for reducing radiation dose in CT. *Radiologic Clinics of North America* **47**, 27–40, doi:10.1016/j.rcl.2008.10.006 (2009).
14. Montgomery, T., Karl, W. C., Castañón, D. A. "Performance estimation for threat detection in CT systems," *Proc. SPIE 10187, Anomaly Detection and Imaging with X-Rays (ADIX) II*, 10187M (1 May 2017).
15. Egan, C. K. et al. 3D chemical imaging in the laboratory by hyperspectral X-ray computed tomography. *Scientific Reports* **5**, 15979, doi:10.1038/srep15979. www.nature.com/articles/srep15979-supplementary-information (2015).
16. de Jonge, M. D. & Vogt, S. Hard X-ray fluorescence tomography—An emerging tool for structural visualization. *Current Opinion in Structural Biology* **20**, 606–614, doi:10.1016/j.sbi.2010.09.002 (2010).
17. Bunge, H. J. in *Texture Analysis in Materials Science* 42–46. (Butterworth-Heinemann, 1982).
18. Chandler, D. in *Introduction to Modern Statistical Mechanics*. (Oxford University Press, 1987).
19. Harding, G. & Delfs, J. in *Optical Engineering + Applications*. 12 (SPIE).
20. Ghammraoui, B. et al. Effect of grain size on stability of X-ray diffraction patterns used for threat detection. *Nuclear Instruments and Methods in Physics Research Section A: Accelerators, Spectrometers, Detectors and Associated Equipment* **683**, 1–7, doi:10.1016/j.nima.2012.04.034 (2012).
21. Engler, O. & Randle, V. *Introduction to Texture Analysis: Macrotexture, Microtexture, and Orientation Mapping*, Second Edition. (CRC Press, 2009).
22. Kocks, U. F., Tomé, C. N., Wenk, H. R., Beaudoin, A. J. & Mecking, H. *Texture and Anisotropy: Preferred Orientations in Polycrystals and Their Effect on Materials Properties*. (Cambridge University Press, 2000).
23. Greenberg, J. A., MacGibbon, C., Hazineh, D., Keohane, B. & Wolter, S. in *SPIE Defense + Security* Vol. 10632. 9 (SPIE, 2018).
24. Greenberg, J. A. & Brady, D. J. in *SPIE Defense + Security*. 8 (SPIE).
25. Yun, W. B., Viccaro, P. J., Lai, B. & Chrzas, J. Coherent hard X-ray focusing optics and applications. *Review of Scientific Instruments* **63**, 582–585, doi:10.1063/1.1142711 (1992).
26. Schneider, S. M., Schlomka, J.-P. & Harding, G. L. in *Medical Imaging 2001*. 10 (SPIE).
27. Schlomka, J.-P., Harding, A., Stevendaal, U. v.,Grass, M. & Harding, G. L. in *Medical Imaging 2003*. 10 (SPIE).
28. Pang, S., Zhu, Z., Wang, G. & Cong, W. Small-angle scatter tomography with a photon-counting detector array. *Physics in Medicine and Biology* **61**, 3734–3748, doi:10.1088/0031-9155/61/10/3734 (2016).

29. Artioli, G. et al. X-ray diffraction microtomography (XRD-CT), a novel tool for non-invasive mapping of phase development in cement materials. *Analytical and Bioanalytical Chemistry* **397**, 2131–2136, doi:10.1007/s00216-010-3649-0 (2010).

30. Cui, C., Jorgensen, S. M., Eaker, D. R. & Ritman, E. L. Direct three-dimensional coherently scattered X-ray microtomography. *Medical Physics* **37**, 6317–6322, doi:10.1118/1.3517194 (2010).

31. Tunna, L. et al. The manufacture of a very high precision X-ray collimator array for rapid tomographic energy dispersive diffraction imaging (TEDDI). *Measurement Science and Technology* **17**, 1767 (2006).

32. *XRD 3500*, www.morpho.com/sites/morpho/files/morpho_xrd3500_dat_r001_0. pdf.

33. Kosciesza, D. et al. in *2013 IEEE Nuclear Science Symposium and Medical Imaging Conference (2013 NSS/MIC)*. 1–5.

34. Shanks, N. E. L. & Bradley, A. L. W. *Handbook of Checked Baggage Screening: Advanced Airport Security Operation*. (John Wiley & Sons, 2005).

35. Harding, G. X-ray diffraction imaging—A multi-generational perspective. *Applied Radiation and Isotopes* **67**, 287–295, doi:10.1016/j.apradiso.2008.08.006 (2009).

36. Harding, G. et al. X-ray diffraction imaging with the Multiple Inverse Fan Beam topology: Principles, performance and potential for security screening. *Applied Radiation and Isotopes* **70**, 1228–1237, doi:10.1016/j.apradiso.2011.12.015 (2012).

37. Sridhar, V., Kisner, S. J., Skatter, S. & Bouman, C. A. "Model-based reconstruction for X-ray diffraction imaging," *Proc. SPIE 9847, Anomaly Detection and Imaging with X-Rays (ADIX)*, 98470K (12 May 2016).

38. Skatter, S., Fritsch, S. & Schlomka, J.-P. "Detecting liquid threats with X-ray diffraction imaging (XDi) using a hybrid approach to navigate trade-offs between photon count statistics and spatial resolution," *Proc. SPIE 9847, Anomaly Detection and Imaging with X-Rays (ADIX)*, 984704 (12 May 2016).

39. Dicken, A. et al. Combined X-ray diffraction and kinetic depth effect imaging. *Optics express* **19**, 6406–6413, doi:10.1364/OE.19.006406 (2011).

40. Evans, P., Rogers, K., Dicken, A., Godber, S. & Prokopiou, D. X-ray diffraction tomography employing an annular beam. *Optics express* **22**, 11930–11944, doi:10.1364/OE.22.011930 (2014).

41. Dicken, A., Shevchuk, A., Rogers, K., Godber, S. & Evans, P. High energy transmission annular beam X-ray diffraction. *Optics express* **23**, 6304–6312, doi:10.1364/OE.23.006304 (2015).

42. Dicken, A. J. et al. Energy-dispersive X-ray diffraction using an annular beam. *Optics express* **23**, 13443–13454, doi:10.1364/OE.23.013443 (2015).

43. Dicken, A. J. et al. Depth resolved snapshot energy-dispersive X-ray diffraction using a conical shell beam. *Optics express* **25**, 21321–21328, doi:10.1364/OE.25.021321 (2017).

44. Gehm, M. E. & Kinast, J. "Adaptive spectroscopy: towards adaptive spectral imaging," *Proc. SPIE 6978, Visual Information Processing XVII*, 69780I (1 April 2008).

45. Golay, M. J. E. Multi-Slit Spectrometry*. *Journal of the Optical Society of America* **39**, 437–444, doi:10.1364/JOSA.39.000437 (1949).

46. Fellgett, P. B. On the ultimate sensitivity and practical performance of radiation detectors. *Journal of the Optical Society of America* **39**, 970–976, doi:10.1364/JOSA.39.000970 (1949).

47. Harwit, M. & Sloane, N. J. A. in *Hadamard Transform Optics* (eds M. Harwit & N. J. A. Sloane) 44–95. (Academic Press, 1979).

48. Gehm, M. E., John, R., Brady, D. J., Willett, R. M. & Schulz, T. J. Single-shot compressive spectral imaging with a dual-disperser architecture. *Optics express* **15**, 14013–14027, doi:10.1364/OE.15.014013 (2007).

49. Tsai, T.-H. & Brady, D. J. Coded aperture snapshot spectral polarization imaging. *Applied Optics* **52**, 2153–2161, doi:10.1364/AO.52.002153 (2013).

50. Llull, P. et al. Coded aperture compressive temporal imaging. *Optics express* **21**, 10526–10545, doi:10.1364/OE.21.010526 (2013).

51. Tsai, T.-H., Llull, P., Yuan, X., Carin, L. & Brady, D. J. Spectral-temporal compressive imaging. *Optics Letters* **40**, 4054–4057, doi:10.1364/OL.40.004054 (2015).

52. Mrozack, A., Marks, D. L. & Brady, D. J. Coded aperture spectroscopy with denoising through sparsity. *Optics express* **20**, 2297–2309, doi:10.1364/OE.20.002297 (2012).

53. Young, L. M. a. N. O. in *Optical Instruments and Techniques* (ed K. J. Habell) 305.

54. Ables, J. G. Fourier transform photography: A new method for X-ray astronomy. *Publications of the Astronomical Society of Australia* **1**, 172–173, doi:10.1017/S1323358000011292 (2016).

55. Dicke, R. H. Scatter-hole cameras for X-rays and gamma rays. *The Astrophysical Journal* **153**, L101 (1968).

56. Hassan, M., Greenberg, J. A., Odinaka, I. & Brady, D. J. Snapshot fan beam coded aperture coherent scatter tomography. *Optics express* **24**, 18277–18289, doi:10.1364/OE.24.018277 (2016).

57. Fenimore, E. E. & Cannon, T. M. Coded aperture imaging with uniformly redundant arrays. *Applied Optics* **17**, 337–347, doi:10.1364/AO.17.000337 (1978).

58. Gottesman, S. R. & Fenimore, E. E. New family of binary arrays for coded aperture imaging. *Applied Optics* **28**, 4344–4352, doi:10.1364/AO.28.004344 (1989).

59. Brady, D. J., Marks, D. L., MacCabe, K. P. & O'Sullivan, J. A. Coded apertures for X-ray scatter imaging. *Applied Optics* **52**, 7745–7754, doi:10.1364/AO.52.007745 (2013).

60. Holmgren, A. *Coding Strategies for X-ray Tomography*, Ph.D. thesis, Duke University, (2016).

61. Rossman, A. et al. in *SPIE Medical Imaging*. 6 (SPIE).

62. FitzGerald, J. *A Programmable Liquid Collimator for Both Coded Aperture Adaptive Imaging and Multiplexed Compton Scatter Tomography*, Master's thesis, AIR FORCE INST OF TECH (2012).

63. Harding, G. Compact multi-focus X-ray source, X-ray diffraction imaging system, and method for fabricating compact multi-focus X-ray source, US patent (2009).

64. Hemberg, O., Otendal, M. & Hertz, H. M. Liquid-metal-jet anode electron-impact X-ray source. *Applied Physics Letters* **83**, 1483–1485, doi:10.1063/1.1602157 (2003).

65. Greenberg, J. A., Hassan, M., Krishnamurthy, K. & Brady, D. Structured illumination for tomographic X-ray diffraction imaging. *Analyst* **139**, 709–713, doi:10.1039/C3AN01641B (2014).

66. Greenberg, J., Iniewski, K. & Brady, D., "CZT detector modeling for coded aperture X-ray diffraction imaging applications," *2014 IEEE Nuclear Science Symposium and Medical Imaging Conference (NSS/MIC)*, Seattle, WA, 2014, pp. 1–3.

67. Veale, M. C. et al. in *IEEE Nuclear Science Symposium & Medical Imaging Conference*. 3789–3792.
68. Tabary, J., Paulus, C., Montémont, G., Verger, L. "Impact of sub-pixelation within CdZnTe detectors for X-ray diffraction imaging systems," *Proc. SPIE 10187, Anomaly Detection and Imaging with X-Rays (ADIX) II*, 101870J (1 May 2017).
69. Iniewski, K., Chen, H., Bindley, G., Kuvvetli, I. & Jorgensen, C. B. in *2007 IEEE Nuclear Science Symposium Conference Record*. 4608–4611.
70. Hassan, M., Holmgren, A., Greenberg, J. A., Odinaka, I. & Brady, D. in *SPIE Defense + Security*. 8 (SPIE).
71. Greenberg, J. A. & Brady, D. J. in *IS&T/SPIE Electronic Imaging*. 11 (SPIE).
72. Lin, Y. et al. in *SPIE Defense + Security*. 6 (SPIE).
73. Coccarelli, D. et al. in *SPIE Defense + Security*. 6 (SPIE).
74. Zhou, M. et al. Nonparametric Bayesian dictionary learning for analysis of noisy and incomplete images. *IEEE Transactions on Image Processing* **21**, 130–144, doi:10.1109/TIP.2011.2160072 (2012).
75. Tikhonov, A. N. & Arsenin, V. I. A. *Solutions of Ill-Posed Problems*. (Winston, 1977).
76. Rudin, L. I., Osher, S. & Fatemi, E. Nonlinear total variation based noise removal algorithms. *Phys. D* **60**, 259–268, doi:10.1016/0167-2789(92)90242-f (1992).
77. Stable signal recovery from incomplete and inaccurate measurements. *Communications on Pure and Applied Mathematics* **59**, 1207–1223, doi:10.1002/cpa.20124 (2006we).
78. Bioucas-Dias, J. M. & Figueiredo, M. A. T. A new TwIST: Two-step iterative shrinkage/thresholding algorithms for image restoration. *IEEE Transactions on Image Processing* **16**, 2992–3004, doi:10.1109/TIP.2007.909319 (2007).
79. Odinaka, I. et al. in *2015 IEEE Nuclear Science Symposium and Medical Imaging Conference (NSS/MIC)*. 1–4.
80. Wang, L. et al. Signal recovery and system calibration from multiple compressive Poisson measurements. *SIAM Journal on Imaging Sciences* **8**, 1923–1954, doi:10.1137/140998779 (2015).
81. Odinaka, I. et al. Joint system and algorithm design for computationally efficient fan beam coded aperture X-ray coherent scatter imaging. *IEEE Transactions on Computational Imaging* **3**, 506–521, doi:10.1109/TCI.2017.2721742 (2017).
82. Odinaka, I., Greenberg, J. A., Kaganovsky, Y., Holmgren, A., Hassan, M., Politte, D. G., O'Sullivan, J. A., Carin, L. & Brady, D. J. "Coded aperture X-ray diffraction imaging with transmission computed tomography side-information," in *SPIE Medical Imaging*, (SPIE, 2016), 9.
83. Zhang, L. & YangDai, T. Determination of liquid's molecular interference function based on X-ray diffraction and dual-energy CT in security screening. *Applied Radiation and Isotopes* **114**, 179–187, doi:10.1016/j.apradiso.2016.05.019 (2016).
84. YangDai, T. & Zhang, L. Spectral unmixing method for multi-pixel energy dispersive X-ray diffraction systems. *Applied Optics* **56**, 907–915, doi:10.1364/AO.56.000907 (2017).
85. Zhao, B., Wolter, S. & Greenberg, J. A. "Application of machine learning to X-ray diffraction-based classification," in *SPIE Defense + Security*, (SPIE, 2018), 6.
86. MacCabe, K. *X-ray scatter tomography using coded apertures*, Doctor of Philosophy thesis, University of North Carolina, (2014).

87. Greenberg, J. A., Lakshmanan, M. N., Brady, D. J. & Kapadia, A. J. "Optimization of a coded aperture coherent scatter spectral imaging system for medical imaging," *Proc. SPIE 9412, Medical Imaging 2015: Physics of Medical Imaging*, 94125E (18 March 2015).

88. Greenberg, J., Krishnamurthy, K. & Brady, D. Compressive single-pixel snapshot X-ray diffraction imaging. *Optics Letters* 39, 111–114, doi:10.1364/OL.39.000111 (2014).

89. MacCabe, K. P., Holmgren, A. D., Tornai, M. P. & Brady, D. J. Snapshot 2D tomography via coded aperture X-ray scatter imaging. *Applied Optics* 52, 4582–4589, doi:10.1364/AO.52.004582 (2013).

90. Greenberg, J. A., Hassan, M., Regnerus, B. & Wolter, S. "Design and implementation of a fan beam coded aperture X-ray diffraction tomography system for checkpoint baggage scanning," *Proc. SPIE 10187, Anomaly Detection and Imaging with X-Rays (ADIX) II*, 1018708 (1 May 2017).

91. Holmgren, A. D., Odinaka, I., Greenberg, J. A. & Brady, D. J. "Multi-view coded aperture coherent scatter tomography," *Proc. SPIE 9847, Anomaly Detection and Imaging with X-Rays (ADIX)*, 98470A (12 May 2016).

92. Kaganovsky, Y. et al. Compressed sampling strategies for tomography. *Journal of the Optical Society of America A* 31, 1369–1394, doi:10.1364/JOSAA.31.001369 (2014).

93. Pang, S. et al. Complementary coded apertures for 4-dimensional X-ray coherent scatter imaging. *Optics Express* 22, 22925–22936, doi:10.1364/OE.22.022925 (2014).

94. Schubert, H. & Kuznetsov, A. *Detection of Bulk Explosives Advanced Techniques against Terrorism: Proceedings of the NATO Advanced Research Workshop on Detection of Bulk Explosives Advanced Techniques against Terrorism St. Petersburg, Russia 16–21 June* 2003. (Springer, Netherlands,2012).

95. Harding, G. X-ray scatter tomography for explosives detection. *Radiation Physics and Chemistry* 71, 869–881, doi:10.1016/j.radphyschem.2004.04.111 (2004).

96. Yuan, S., Wolter, S. D. & Greenberg, J. A. "Classification-free threat detection based on material-science-informed clustering," in *SPIE Defense + Security*, (SPIE, 2017), 8.

97. Heimann, S. www.donggok.co.kr/data/HS10065HDX.pdf,www.donggok.co.kr/data/HS10065HDX.pdf (2004).

98. Diallo, S. O. et al. "Towards an X-ray-based coded aperture diffraction system for bulk material identification," in *SPIE Defense + Security*, (SPIE, 2018), 16.

99. Lakshmanan, M. N., Greenberg, J. A., Samei, E., Kapadia, A. J. Design and implementation of coded aperture coherent scatter spectral imaging of cancerous and healthy breast tissue samples. *Journal of Medical Imaging* 3(1), 013505, 2016. doi:10.1117/1.JMI.3.1.013505.

100. Royle, G. J. & Speller, R. D. Quantitative X-ray diffraction analysis of bone and marrow volumes in excised femoral head samples. *Physics in Medicine & Biology* 40, 1487 (1995).

101. Uvarov, V. et al. X-ray diffraction and SEM study of kidney stones in Israel: Quantitative analysis, crystallite size determination, and statistical characterization. *Environmental Geochemistry and Health* 33, 613–622, doi:10.1007/s10653-011-9374-6 (2011).

102. Desouky, O. S., Elshemey, W. M. & Selim, N. S. X-ray scattering signatures of β-thalassemia. *Nuclear Instruments and Methods in Physics Research Section A: Accelerators, Spectrometers, Detectors and Associated Equipment* **607**, 463–469, doi:10.1016/j.nima.2009.05.160 (2009).

103. Carboni, E. et al. Imaging of neuronal tissues by X-ray diffraction and X-ray fluorescence microscopy: Evaluation of contrast and biomarkers for neuro-degenerative diseases. *Biomedical Optics Express* **8**, 4331–4347, doi:10.1364/ BOE.8.004331 (2017).

104. Scott, R., Stone, N., Kendall, C., Geraki, K. & Rogers, K. Relationships between pathology and crystal structure in breast calcifications: An in situ X-ray diffraction study in histological sections. *Npj Breast Cancer* **2**, 16029, doi:10.1038/npjbcancer.2016.29. www.nature.com/articles/npjbcancer201629-supplementary-information (2016).

105. Moss, R. M. et al. Correlation of X-ray diffraction signatures of breast tissue and their histopathological classification. *Scientific Reports* **7**, 12998, doi:10.1038/ s41598-017-13399-9 (2017).

106. Kern, K., Peerzada, L., Hassan, L. & MacDonald, C. "Design for a coherent-scatter imaging system compatible with screening mammography," in (SPIE, 2016), 4.

107. Theodorakou, C. & Farquharson, M. J. Human soft tissue analysis using X-ray or gamma-ray techniques. *Physics in Medicine & Biology* **53**, R111 (2008).

108. Chrysoula, T. & Michael, J. F. The classification of secondary colorectal liver cancer in human biopsy samples using angular dispersive X-ray diffraction and multivariate analysis. *Physics in Medicine & Biology* **54**, 4945 (2009).

109. Drakos, I., Kenny, P., Fearn, T. & Speller, R. Multivariate analysis of energy dispersive X-ray diffraction data for the detection of illicit drugs in border control. *Crime Science* **6**, 1, doi:10.1186/s40163-016-0062-9 (2017).

110. Lee, D. C. & Webb, M. L. *Pharmaceutical Analysis.* (Wiley, 2009).

111. Greenberg, J. *Detecting the presence of copper nanoparticles in wood via coded aperture X-ray diffraction tomography* (2015).

2

Semiconductor Sensors for XRD Imaging

Krzysztof Iniewski and Adam Grosser

Redlen Technologies

CONTENTS

2.1 Indirect vs. Direct Conversion Technology

2.1.1 Introduction

There are two ways to detect X-ray radiation today: indirect conversion using scintillator materials and direct conversion using semiconductor sensors. The main difference between an indirect conversion scintillator detector and a direct conversion semiconductor detector can be summarized as follows. The scintillation-based indirect detector is an optical device using light photons as intermittent information carriers. A direct conversion detector omits the need for the conversion of radiation to light and directly generates charge carriers (electrons) from the energy of the absorbed X-ray photon. It is an electrical device that employs electrons and holes to transfer the event information to the electrodes. The difference between two types of detection is schematically summarized in Figure 2.1. Details of the comparison will be discussed later in this chapter.

In general, radiation detectors can be operated in an integrating or a counting mode. In integrating mode, the charge information is sampled over an integration time and converted to a digital signal. In counting mode, the process in direct conversion detectors, the total number of events is measured by counting the charge pulses as illustrated in Figure 2.2. In addition, the energy of each absorbed photon can be obtained by measuring the total charge or pulse amplitude of each photon. Direct conversion detector thus offers spectral resolution of the incoming radiation. The additional spectral

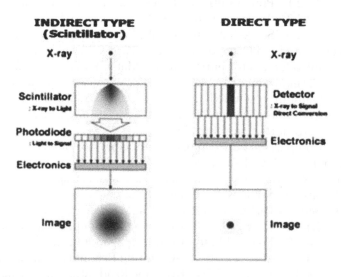

FIGURE 2.1
Schematic comparison between conventional detector using scintillator (left) and direct conversion detector (right) [1].

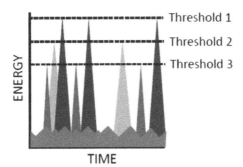

FIGURE 2.2
Schematic illustration of the photon-counting principle. Multiple thresholds can be used to classify the incoming photon energy.

information leads to well-recognized benefits of the photon-counting (PC) technology: lower noise, higher efficiency and better spatial resolution. X-ray imaging applications benefit from direct conversion as it requires less image filtering for the same obtained image resolution or better image resolution at the same amount of image filtering.

Currently, most digital radiation detectors are based on integrating the X-ray photons emitted from the X-ray tube for each frame. This technique is vulnerable to noise due to variations in the magnitude of the electric charge generated per X-ray photon. Higher energy photons deposit more charge in the detector than lower energy photons so that in the integrating detector, the higher energy photons receive greater weight. This effect is undesirable as the higher part of the energy spectrum provides lower differential attenuation between various materials or tissues, and hence, these energies yield images of low contrast. X-ray counting detectors solve the noise problem associated with photon weighting by providing better weighting of information from X-ray photons with different energies. In an X-ray quantum counting system, all photons detected with energies above a certain predetermined noise threshold are assigned as a signal. Adding energy ranges, referred to as "'energy binning", into the system (i.e., counting photons within a specified energy range) theoretically eliminates the noise associated with photon weighting and decreases the required X-ray dosage by up to 50% compared to the integrating systems.

2.1.2 Indirect Conversion Using Scintillator Materials

Scintillator materials are used to detect and convert incoming X-rays into visible light. Upon receiving X-ray photons, scintillators capture their energy and release it as photons of lower energy. Is it desirable that the material should capture as much of the energy of each photon striking it as possible in order to build an accurate photon energy spectrum and thus identify the material emitting the photons. Ideally, a photon should deposit its full energy in the scintillator material, a so-called full energy interaction. It is also important that the deposited energy be efficiently converted into photons of visible light, which are then counted to determine the original X-ray photon energy.

An example of the detection process using scintillator material is shown in Figure 2.3. An X-ray photon of energy E might be absorbed by the scintillator, which is the desired process in the X-ray detection. However, an X-ray photon can also scatter and be detected in a neighboring piece of the scintillator detector or escape the detector volume completely. After absorption, multiple visible photons are generated and propagate in all directions, irrespective of the direction of the original photon. This leads to significant loss of instrument efficiency. Finally, the visible photons are typically detected by standard silicon photodiodes.

2.1.3 Director Conversion Using Semiconductor Materials

The direct detection process – used by semiconductor materials such as Silicon (Si), Germanium (Ge), Cadmium Telluride (CdTe) and Cadmium Zinc Telluride (CZT) sensors – is illustrated in Figure 2.4. It offers better efficiency and resolution compared to the traditional method using scintillators described earlier and is ideally suited for XRD diffraction application.

When a high-energy photon strikes a CZT detector, hole-electron pairs are generated within the material. The charge generated is proportional to the photon energy and given by the following equation [3]:

$$Q_0 = q \left(\frac{E_{\text{photon}}}{E_{\text{pair}}} \right)$$

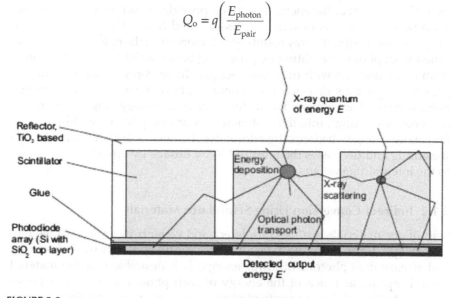

FIGURE 2.3
Illustration of the two-step detection process using scintillator and silicon photodiode [2].

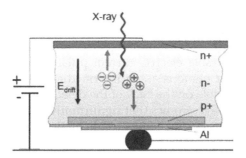

FIGURE 2.4
Photon to charge conversion in direct conversion sensors.

Q_o is the charge generated in Coulombs, q is the charge of an electron, E_{photon} is the energy of the incident photon in keV, and E_{pair} is the ionization energy required to generate an electron-hole pair in the sensor material.

For diffraction and imaging applications, the direct conversion detectors are typically pixelated. When a photon strikes a certain pixel, a small charge is generated that must be sensed and processed. Over the past decade, this is often done with an Application-Specific Integrated Circuit (ASIC) that is directly attached to the CZT detector. Modern integrated circuit technology makes it economically feasible for each pixel to have its own low-noise electronics, creating what are known as hybrid pixel detector readout ASICs. The block diagram of the electronics for each pixel or channel is shown below in Figure 2.5. Chapter 3 of this book is entirely devoted to issues related to the readout ASIC electronics.

Direct conversion radiation detectors offer new capabilities for medical imaging, NDT and security applications due to their superior energy resolution (ER), detector quantum efficiency (DQE), variable energy weighting, noise reduction as well as increased spatial resolution. As the result of their superior ER, direct conversion detectors are the only devices available for XRD applications. Their advantages in turn enable new applications such

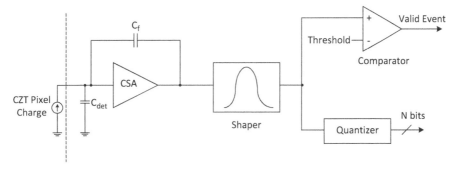

FIGURE 2.5
Charge detection electronics block diagram.

as material decomposition as well as improved image quality in low-dose screening applications and high spatial resolution imaging. In direct conversion detectors, absorbed X-ray imaging photons generate individually detectable and measurable electric current pulses without the production and detection of light photons as an intermediary step. Since the signals are immediately available on the back of the detectors, they are sensed with close fitting miniature electronics that significantly reduces sources of noise and expensive infrastructure overhead. Schematic illustration of the direct detection architecture is shown in Figure 2.6.

Various semiconductor materials like silicon (Si), germanium (Ge), amorphous Selenium (a-Se), Gallium Arsenide (GaAs), Cadmium Telluride (CdTe) and Cadmium Zinc Telluride (CZT) can be used in direct detection. The comparison of different detector systems shows that semiconductor materials like CdTe/CZT are excellent material choices for

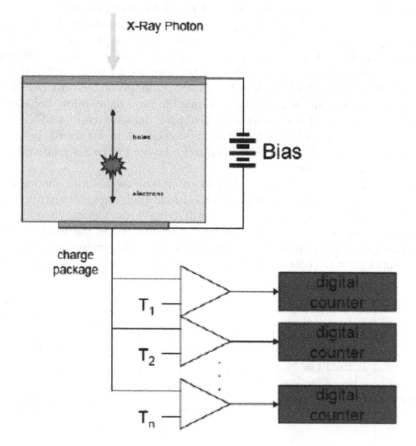

FIGURE 2.6
Schematic representation of the direct readout process that uses semiconductor detector and photon-counting electronics.

high-performing PC X-ray detectors for high X-ray energies. The energy resolution (ER) expressed as Full-Width Half Max (FWHM) value is superior to the common scintillators like NaI and CsI (caesium iodide) as shown in Figure 2.7.

In addition, CdTe/CZT carrier trapping (expressed as carrier mobility-lifetime product) is superior to that of amorphous Selenium (a-Se), allowing for fast detectors free from image ghosting (residual images left over from previous scan) and lag that is typical for scintillator materials. Finally, the photon absorption efficiency is very high.

High levels of energy discrimination and PC binning enable material determination in CdTe/CZT systems, in contrast to conventional integrating detectors that do not offer such opportunities. The dynamic range measurements on the CdTe/CZT detectors illustrate one of the key benefits of our photon-counting technology; the detector is free from electronic noise and has excellent dynamic range, from zero up to hundreds of millions of photons/pixel/s. In conclusion, the photon-counting CdTe/CZT detectors offer unique performance with excellent spatial resolution, no electronic noise, high quantum efficiency, high frame rates and an extremely wide dynamic range. This, in combination with the electronic energy separation, has the potential to improve present and enable new applications in medical imaging, security, NDT and related fields.

Despite intense efforts on the part of established scintillator suppliers, none of the materials available in the marketplace can compete with CZT sensors. As discussed above, CZT offers simple direct conversion process that does not require additional X-ray-to-visible light conversion and the use of photo multiplying tubes (PMTs) or avalanche photo diodes (APDs). CZT produces 20,000 electrons for 100 keV photon vs. only about 1,000 for Thallium doped sodium iodide NaI(TI), resulting in 20× enhancement in detection efficiency.

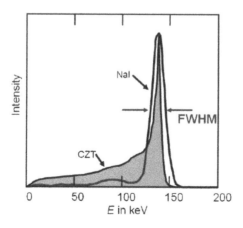

FIGURE 2.7
Comparison between CZT and NaI scintillator illustrating better energy resolution [2].

Finally, CZT provides better ER, as shown in Figure 2.7, which enables clear identification of various materials, tissues or radioisotopes – a critical feature in NDT, medical and security applications.

2.1.4 Application Example: Computed Tomography

In computed tomography (CT), traditional scanners use indirect conversion with scintillation detectors like gadolinium oxysulfide (GOS) or caesium iodide (CsI), where each pixel is enclosed by an epoxy compound filled with back-scattering particles. Typical pixel dimensions of around 1 millimeter are typically used. A registered photo sensor detects the secondary light photons at the bottom surface of each pixel. The primary interaction in a detector pixel is given by the absorption of an incoming X-ray quantum by a gadolinium atom (in the case of GOS). The X-ray energy is converted into light photons. The energy conversion rate is around 12%. Secondary light photon transport takes place and the photons which reach the photosensor contribute to the output energy signal.

Radiography and mammography detectors follow similar designs. CsI is usually employed as a scintillator. Due to its vertical needle structure, it has the advantage of providing intrinsic light guiding properties; thus, no back-scattering septa are required. This allows for an improved detector resolution at the expense of a reduced stopping power and signal speed.

Two main physical effects influence the spatial and spectral resolution in pixelized scintillator detectors. First, the primary energy deposition is not perfectly localized. For the high-Z atom gadolinium, absorption is governed by the photoelectric effect. This generates fluorescence escape photons with mean free path lengths in the order of several hundreds of microns. They might be reabsorbed in the pixel volume, become registered in a neighboring pixel or leave the detector volume completely. Second, light transport is affected by optical cross-talk. Septa walls are designed with a limited thickness to optimize overall dose usage and light yield. As a consequence, a significant portion of the light is transferred to adjacent 'false' pixels, creating a blurry image in the process.

The direct conversion scheme does not have these disadvantages. A common-cathode design with pixelized anodes on the bottom surface of the semiconductor bulk is typically used. Imaging voxels are established with electrical fields in the order of 1 kV/cm. The physics of the primary energy deposition is comparable to the indirect conversion detector. However, the deposited energy is converted to charges instead of optical photons. The holes and electrons are separated and accelerated by the electrical field. Electrical pulse signals are induced on the electrodes, and the main signal pulse is generated when the electrons follow the stronger curved electrical field in the bulk region close to the anodes.

Direct detection based CT readout systems utilize multiple bin PC where the radiologist can set various thresholds (energy bins). In photon counting, only pulses with heights higher than thresholds are counted, and the basic operation is noise-free because the lowest threshold can be set above the noise level. The number of energy bins varies from 2 to 6, with 4 being typical. The ability to count photons within separate energy bins enables the use of some innovating energy weighting scenarios in image reconstruction.

2.1.5 Application Example: Baggage Scanning

Baggage scanning equipment needs to detect explosives in bulk, sheet, liquid and slurry form. Traditional radiation detection systems use scintillator technology, as shown in Figure 2.8. X-rays produced by the X-ray tube are used to scan the object. Ceramic scintillator detects those photons that have passed through the object and produce a visible light signal. The received light signal is, in turn, converted by the photodiode to produce analog electrical signals that can be used to produce an image.

Most detection systems take linear projections through the luggage travelling on a conveyor belt in a so-called line scan mode. The detectors used are efficiently collimated linear arrays. For dual-energy capabilities, two solutions have been used. In the first technique, the linear scan is performed twice, once without, and then repeated with an additional X-ray filter in front of the beam. In this way, an elementary technique of low- and high-energy photon separation can be obtained. Similarly, the first scan can be performed at 1 kVp tube voltage (e.g., 140 kVp) and the second at another (e.g., at 80 keVp).

In the second technique, the linear scan is performed only once but sandwich detectors have been optimized for dual energy scanning. They consist of two layers of scintillator-photodiode type, separated by a metal filter. The first layer absorbs low energy and the second layer absorbs the high-energy photons.

In both cases, due to the poor energy separation of those acquisition systems and to a significant noise level resulting from the acquisition speed, the obtained accuracy at best only allows materials to be classified into broadbands, such inorganic and organic.

Modern baggage scanning equipment uses more efficient, direct method of radiation detection, as shown in Figure 2.9. The two-step process involving

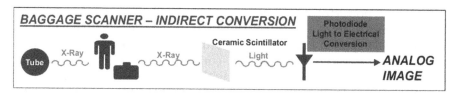

FIGURE 2.8
Scintillator-based baggage scanning technology as typically used in standard systems.

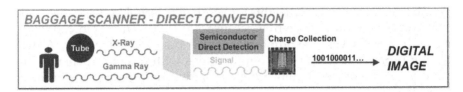

FIGURE 2.9
Semiconductor-based baggage scanning technology being introduced today.

scintillator and photodiode is replaced by a semiconductor detector that converts X-rays directly into electric charge. High-purity Germanium (HPGe) is currently used in commercial systems, but CdTe and CZT sensors have recently attracted attention. CZT detectors in particular fit perfectly into these applications due their high stopping power, reasonable cost, high stability and reliability.

In addition, through the use of CZT, multiple energy analysis and multiple energy sources, the equipment might be able to detect thin sections and subtle differences in atomic number. The equipment might also provide full volumetric reconstruction and analysis of the object being imaged. The result of the advanced CZT technology could be the exceptional ability to detect and discriminate a wide variety of threat materials, providing enhanced overall security. One of the key motivations in technological developments is to eliminate the ban on carrying liquid that is currently in place. Once this ban is eliminated (already legislated in Europe), the airport operators will be forced to use better technology for liquid/security detection.

An example of such a technology is shown in Figure 2.10. The suitcase enters the device at the right of figure where spatial landmarks for registration purposes are measured by pre-scanner. In the main housing (center-left of Figure 2.10), a primary cone-beam executes a meander scan, either of a region-of-interest or suitcase in its entirety. The underlying principle of detection is X-ray diffraction imaging (or XRD), which is the subject of this book. XRD tomography (XRDT) refers to the volumetric analysis of extended, in homogenous objects by spatially resolved X-ray diffraction and is discussed in more detail later in the following chapters. In particular, the X-ray diffraction technology is discussed in Chapter 7 entitled "Energy resolving detectors for XDi airport security systems".

XRD is a powerful analytical tool that has been used for the non-destructive analysis of a wide variety of materials for nearly 100 years. XRD is now widely applied in a variety of industries including metallurgy, photovoltaics, forensics, pharmaceuticals, semiconductors and catalysis and can be used to analyze virtually any solid with crystalline structure. It has been recently applied to security detection. The fundamental strength of the technique is its ability to characterize the periodic atomic structure present

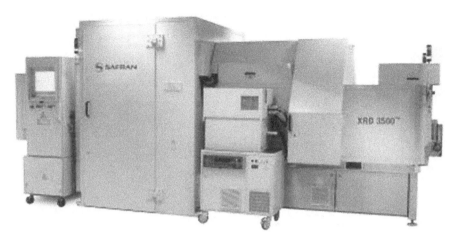

FIGURE 2.10
Direct tomographic, energy-dispersive XRD 3500 system from Morpho-Detection (www. morpho.com).

in crystalline or polycrystalline materials. It also has some capability to distinguish liquids.

In XRD analysis, a sample is illuminated by a collimated X-ray beam of known wavelength. If the material is crystalline, it possesses a 3D ordering or "structure" with repeat units of atomic arrangement (unit cells). X-rays are elastically scattered (i.e., diffracted) by the repeating crystal planes lattice of materials, while X-rays are more diffusely scattered by amorphous materials. X-ray diffraction occurs at specific angles with respect to the lattice spacings defined by Bragg's Law. Any change or difference in lattice spacing results in a corresponding shift in the diffraction lines. It is this principle that such properties as identification (based on phase) and residual stresses are obtained.

The application of XRD in security detection provides a very powerful opportunity to detect chemical/structural property of the material under investigation. For effective baggage screening, this technique is frequently coupled with CT to visualize the object in question. Traditionally, this is done in a serial fashion, therefore making the scan time long and equipment expensive. With direct conversion detectors, it is possible to perform CT-like imaging and XRD-like detection simultaneously.

A more detailed illustration of the commercial application of the XRD technology is shown in Figure 2.11. Note the difference obtained in scattered X-ray spectra for Semtex (explosive) and some non-hazardous materials. The HPGe detector used in the XRD 3500™ system can be substituted by CZT, which offers similar ER (a few percent over the 30–120 keV energy range). However, none of the scintillator materials can be used for this application, as their ER is too poor.

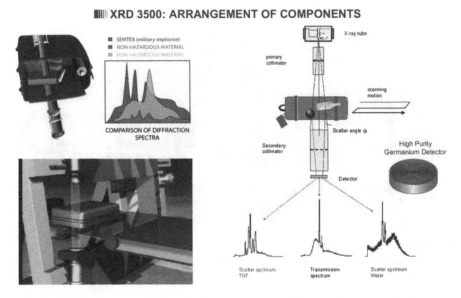

FIGURE 2.11
XRD 3500 architecture and principle of operation (www.morpho.com).

2.2 Scintillator Materials and Photo-Diodes Technologies

2.2.1 Scintillator Materials

2.2.1.1 Inorganic Scintillators

There are two main types of scintillators, inorganic and organic. Inorganic scintillators (those not containing carbon) are typically single crystals, such as sodium iodide (NaI) with a small amount of thallium (Th) added. The probability of full energy interaction increases sharply with atomic number (Z) of the scintillator material and is high for inorganic crystals. The more energy from each photon, a scintillator absorbs and then re-emits, the better the correlation between the energy input and output and the more precise the spectrum that can be constructed.

Inorganic scintillators are quite inexpensive but have several disadvantages. The materials are typically fragile. Many inorganic crystals absorb water and are sensitive to light, so they must be protected from environmental conditions. NaI crystals are easy and inexpensive to grow, but other, higher resolution scintillators are harder to grow, are more costly, and can be even more sensitive to moisture. For NaI, the light output varies strongly with temperature, so the temperature must be stabilized or the data corrected.

2.2.1.2 Organic Scintillators

Organic scintillators have the opposite set of properties. They can be made of plastic, such as PVT (polyvinyltoluene). As such, they are easy and cheap to make and are much less fragile than crystals. They can be produced in bulk, making them suitable for deployment in large sheets, such as for radiation portal monitors. On the other hand, since they are composed mostly of hydrogen and carbon, both very low Z elements, they are very inefficient at absorbing the total energy of gamma rays.

2.2.1.3 Technology Limitations

Regardless of whether inorganic or organic, all scintillators suffer from a major fundamental limitation called *afterglow*, since scintillation crystals contain a number of luminescent components. The main component corresponds to the decay time; however, less intense and slower components also exist. Commonly, the strength of these components is estimated by using the intensity of a scintillator's glow, measured at a specified time after the decay time. Afterglow is the ratio of the intensity measured at this specified time to the intensity of the main component measured at the decay time. Afterglow, which is typically in order of milliseconds, limits the speed with which the X-ray signals can be read using conventional scintillator technology. As discussed in the next section, only direct conversion detectors offer high speeds of operation due to their potential glow-free properties.

2.2.2 Photo-Diode Technologies

Conventional X-ray detection systems are made of three parts: scintillator crystal, photo detectors and a multi-channel digital acquisition system. As discussed above, the role of the scintillator crystal is to absorb X-ray photons and emit visible light, while the photodetection system converts the visible light signal into electric charge or current. When these X-rays interact in the scintillator via the photoelectric effect, electrons are released and travel short distances to luminescent centers in the crystal, where subsequent energy transitions lead to the emission of multiple low energy photons. The energy of these emitted photons varies with the scintillator material but typically these photons are in the visible light spectrum. While the signal produced in the scintillator is proportional to the energy of the detected photon, it is usually quite weak and needs to be amplified with a sensitive photodetector coupled to the crystal. The readout process is summarized in Figure 2.12.

2.2.2.1 Photomultiplier Tubes

The most common device to detect the low intensity light produced in the scintillator used to be the photomultiplier tube (PMT). A PMT is a vacuum

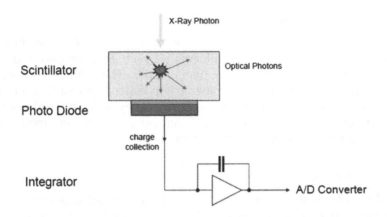

FIGURE 2.12
Schematic representation of indirect readout process that uses scintillator, photo diodes and integrating electronics.

tube containing a light-sensitive photocathode at one end and a series of metal plates called *dynodes* along the length of the tube. At the cathode, visible light from the scintillator is converted into electrons which are then accelerated due to the applied potential difference to the first dynode. For each electron that collides with the dynode, several electrons are released and accelerated towards the next dynode where the interacting electrons further generate more electrons. This process of amplification repeats itself through all dynode stages until eventually the electrons reach the anode. At each dynode stage, for each electron that collides, it is common that between 3 and 6 electrons are released. As most PMTs consist between 8 and 10 dynodes, it is common to achieve an amplification of around a factor of a million through the course of the dynode amplification process.

At the anode, the large number of electrons present creates a measurable electrical signal on the order of several microamps. When passed through an appropriate resistor, voltage drops on the order of several volts that can be obtained. While PMTs have been around for many years and are used extensively in many instruments, they suffer from many disadvantages. For example, they do not operate properly in the vicinity of a magnetic field which precludes their use in MRI applications. In fact, even the earth's relatively weak magnetic field greatly affects the quality of images obtained with conventional PET and SPECT cameras. Care must be taken when orienting conventional PMT-based radiation detectors, and proper calibrations are required. Most SPECT and PET cameras are PMT based and rotate. There is always a metal shield around each tube to prevent influence of magnetic field which increases the unit cost. Additionally, PMTs are known to be extremely sensitive to temperature variations, and their high-operating voltages make them sensitive to noise fluctuations.

2.2.2.2 Avalanche Photodiodes

As PMTs suffer from various limitations outlined above, alternative photo detectors are being used. The workhorse of the photodiode technology is avalanche photodiodes (APD). Avalanche photodiodes essentially consist of a pn-junction semiconductor which, when exposed to light photons from a scintillator crystal, generates electron-hole pairs. When a high reverse bias is applied across the APD diode, an electric field is produced between the anode and cathode and the electron-hole pairs migrate across the depletion layer, holes moving to the cathode and electrons moving to the anode. If the applied bias is high enough, the electrons will be accelerated such that they collide with the surrounding crystal lattice of the semiconductor thereby creating additional ionization effects leading to higher number of additional electron-hole pairs, hence the term "avalanche effect". As the electrons reach the anode, there is a voltage drop in the high voltage bias which is proportional to the original number of electron-hole pairs produced. The signal gain of an APD is proportional to the voltage bias applied across it. There comes a point, however, where further increases to the bias will not cause a proportional voltage drop across the diode. At this voltage, commonly known as the "breakdown voltage", the APD will conduct incident light.

As the voltage drop across an APD tends to be so small, it is beneficial to utilize large active area APDs in order to collect as many light photons as possible and thereby maximize the output voltage signal. When used for radiation detection, most APD detectors typically use a 1:1 coupling of scintillator crystal to APD. This is primarily done because the APD signal is so small that any light sharing among multiple APD detectors as a result of using a monolithic detector would result in a subsequent reduction in the APD signal. This type of arrangement tends to limit the physical size of the detector due to:

- high cost of pixilated scintillators
- difficulty in fabrication
- need for many thousands of digitization channels

For these reasons, APDs are still not necessarily the best choice of photodetectors for many applications. The primary downside is their relatively low signal gain and high noise levels compared to conventional PMTs. While PMTs have typical amplifications of 10E+6, they typically only have gains of about 50–200. This low gain, when combined with a relatively large background noise component, reduces spatial resolution and limits the range of photon energies that can be imaged with such a system. Since the magnitude of the APD signal is dependent upon the number of light photons produced in the scintillator, APDs have typically been limited to high-energy applications. In addition, due to variations in manufacturing, it is important that each APD have its own voltage bias control in order to operate at optimal signal gain. Confounding the issue even further is the fact that APDs exhibit significant gain variations with temperature, thus leading to the requirement of temperature-sensitive operating bias.

2.2.2.3 Solid-State Photomultipliers

In the last few years, a new technology based on avalanche photodiodes has become available. These so-called "solid state photomultipliers", or SSPMs, are similar in many respects to avalanche photodiodes but provide signal amplifications much closer to that of conventional PMTs (up to 10^6). This is achieved by operating the APD in "Geiger mode". As APDs can be made very small (several μm across) and many of these single element APDs placed in an array on a single substrate and all operated in Geiger mode, then by simply counting the number of elements that cascade into Geiger mode, an accurate measure of the original light intensity can be made. Thus, even though each APD element produces a relatively small signal (several mV), the summation of all the elements following a scintillation event yields a rather large, easily detectable signals through the diode, thus causing a much larger than expected voltage drop. At this point, the APD is said to be operating in "Geiger Mode" as the device no longer outputs a voltage proportional to the amount of incident light.

2.3 Direct Conversion Detectors

2.3.1 Introduction to Semiconductor Materials

Silicon is the most commonly used semiconductor sensor material, due to its various advantages. Silicon wafer processing technology is mature as it relies on advancement in a $400 billion (annually) CMOS industry. Silicon wafers are available with near-perfect crystal quality; low-leakage current can be achieved by using a photodiode structure; and silicon sensors are relatively robust, both mechanically and chemically.

However, the X-ray absorption efficiency of silicon is very limited at higher energies. Silicon sensors with a typical thickness of 500 μm provide 90% photoelectric absorption efficiency at 12 keV, but their efficiency falls quickly as the photon energy is increased beyond this value as shown in Figure 2.13. For many applications in X-ray detection, such a low efficiency is not acceptable as they limit the throughput of the system.

Efficient direct detection of hard X-rays can be achieved using the so-called "high-Z" (high atomic number) semiconductors such as Ge, GaAs, CdTe and CZT. Each of these high-Z materials has its own strengths and weaknesses.

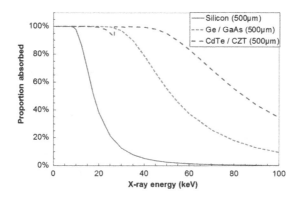

FIGURE 2.13
Photoelectric absorption efficiency of Si, Ge/GaAs and CdTe/CZT.

A common issue for these materials (with the exception of Ge) is that they cannot be produced with the same level of crystal perfection and uniformity as silicon. This is largely due to these materials being compound semiconductors, meaning they contain more than one component which makes their crystal growth more challenging.

2.3.2 Germanium

Unlike most high-Z semiconductors, Ge is an elemental semiconductor, which is a great advantage for producing high-quality crystals. High-purity germanium (HPGe) single crystals are grown through the Czochralski process. Net impurity concentrations can be extremely low (10E+9 cm^{-3}) in both p- and n-type Ge over large volumes. However, because of the relatively low bandgap (0.7 eV), HPGe detectors cannot be operated at room temperature due to the large intrinsic carrier concentration (10E+13 cm^{-3}) leakage current from thermally generated charge carriers. As a result, cooling down to 77 K, the temperature of liquid nitrogen (LN2), is required to reduce the leakage currents to the acceptable level (few pA per pixel). The cooling requirements make this solution quite expensive and difficult to use in commercial applications.

The detection principle of an HPGe semiconductor sensor consists of developing a p–i–n diode structure in which the intrinsic region has high resistivity and is sensitive to ionizing radiation when a reverse-bias voltage is applied. These properties are due to excellent mobility of electron and holes, the relatively high-Z (32) and density (5.32 g/cm^3), and the availability of large, pure and high-quality single crystals. For example, an ER of 0.5 keV can be achieved at 60 keV photon energy. Compared to other semiconductors, HPGe presents the best properties for high-resolution and high-efficiency spectroscopy but very few applications can afford its steep price and cooling requirements.

2.3.3 Gallium Arsenide

Gallium Arsenide (GaAs) is widely used in applications such as high-speed and high-power electronics (GaAs MESFETs) and optoelectronics. The material is available in large wafers (6 inches), which makes it an appealing option for radiation detection. However, detecting X-rays requires a combination of good carrier transport, high resistivity and large sensor thickness (more than 1 mm), which is impossible to achieve today using commercial bulk GaAs produced for other uses in electronics.

In recent years, it has been shown that moderate level of radiation performance can be achieved by taking standard n-type GaAs (which does not have high resistivity) and compensating for excess electrons with chromium (Cr) doping. It is experimentally shown that a material with resistivity of about 10E+9 Ωcm, which is close to the theoretical limit, can be produced using this technology with reasonable charge-transport properties. With a strong electric field applied to the detector, the mean carrier path for electronics is at best of the order of 5–10 mm for electrons but only 50–100 µm for holes; this is acceptable for low-flux applications and not for high-flux applications as NDT. The best ER obtained to date is about 2–3 keV at 60 keV. Poor hole transport precludes the use of GaAs sensors in most commercial applications although its low price can be appealing in some cases where the poor hole transport can be tolerated.

2.3.4 Cadmium Telluride

For applications in NDT, CdTe might be an appealing option because the higher atomic numbers of Cd and Te (48 for Cd, 52 for Te) result in significantly shorter X-ray attenuation lengths compared to Ge, and GaAs 500 µm of CdTe will absorb 55% of 80 keV photons. CdTe is mostly used for applications such as thin-film solar cells, but the detector grade production requires special manufacturing steps. Typically, polycrystalline CdTe ingots are grown from a Te-rich melt, and then the material is progressively recrystallized by the travelling heater method (THM).

The carrier mean free path in CdTe is about 1–10 cm for electrons and 0.1–1 mm for holes, better than GaAs but worse than CZT. Defects in CdTe crystal cause a range of problems, and due to the fragility of the material, some defects such as dislocations can be introduced during processing – for example, mechanical force can cause atomic planes to slip with respect to each other.

CdTe sensor technology has been extensively studied in the last several years [4–15]. One severe issue with CdTe is material polarization, where the electric field within the sensor changes with time. The changes in electric field are due to carrier trapping (mostly holes trapped by deep level acceptors in CdTe). A changing electric field leads to changes in measured radiation spectra making use of these sensors challenging even in low-flux applications.

Under high-flux conditions the polarization effect is much stronger in CdTe due to the amount of holes generated. The trapped charge can build up to a point where the electric field collapses, resulting in complete loss of sensor operation. However, even before catastrophic loss of device operation, serious degradation occurs if the electric field is changing with time.

The effects of polarization depend on the contact technology used too. By using different metals for electrodes, it is possible to create Schottky contacts with the material (giving diode-like behavior) or conductive (ohmic) contacts. Schottky contacts reduce current flow in the sensor, resulting in better spectroscopic performance, but polarization effects increase. Significant signal loss after operating a Schottky contact sensor even for several minutes has been widely reported in the literature. However, this can be combatted by temporarily switching off the high voltage on the sensor and reversing its polarity. Another possibility is to illuminate the sensor with light to create additional hole-electron pairs that would help in carrier recombination process. Needless to say, both solutions lead to major complications in system design. For this reason, Ohmic contacts are more commonly used for high-count-rate applications, but even with use of these contacts, polarization remains a major challenge limiting the thickness of CdTe to typically 1 mm (in some demanding cases like CT scanner, 1.4 mm thickness was used).

2.3.5 Cadmium Zinc Telluride

Replacing some Cd atoms with Zn produces $Cd_{1-x}Zn_xTe$ (CZT) compound that increases the bandgap of the material from 1.44 eV to approximately 1.6 eV for typical Zn concentrations of 10%. The wider bandgap leads to higher sensor resistivity giving CZT a 10× advantage over CdTe. Higher resistivity offers the benefit of lower leakage current, which in turn can improve spectroscopic behavior. Finally, lower leakage current improves detector ER and a result contrast to noise ratio (CNR) in the phantom studies [3].

Superior CZT sensor performance is due to dramatic advances in crystal growth and sensor manufacturing. Different growth methods have been used to grow CZT. The high pressure Bridgman technique produces large, polycrystalline CZT ingots, which can then be diced to obtain single crystals, typically up to a few cubic centimeters. This can provide CZT sensors for spectroscopic detectors, but photon-counting sensors typically require larger single-crystal areas. Recent advances in THM growth have made high-flux applications feasible.

As previously discussed for CdTe, high-flux operation of CZT sensors until recently presented well-known challenges due to poor hole transport and material defects that lead to polarization and other instability effects. Improvements in THM growth and device fabrication that require additional processing steps have enabled CZT to show dramatically improved hole-transport properties and reduced polarization effects. As a result, high-flux operation of CZT sensors at rates in excess of 200 million counts/s/ mm² is now possible. Availability of fast, stable and reliable CZT sensors has enabled multiple medical-imaging original equipment manufacturers to start building cameras for computed tomography, baggage scanning and non-destructive testing (NDT).

The main advantage of the CZT sensor over CdTe is the flexibility in the choice for sensor thickness and the resulting working energy range. As CZT does not suffer from polarization, the CZT sensors can be made significantly thicker. An example of a comparison between 5 mm thick CZT and 1 mm thick CdTe measured at an independent national laboratory in the United Kingdom illustrates the dramatic difference in detector efficiency, as shown in Figure 2.14.

Both sensors have been exposed to a Co-57 source for 10 minutes from a distance of 25 cm. During that exposure, the CZT sensor detected 7145 peak events while the CdTe detected only 795 peak events. The difference of almost an order of magnitude was mostly due to the sensor thickness which was 5 mm for CZT and 1 mm for CdTe. Additional loss of efficiency for CdTe was due to smaller pixel pitch and resulting number of charge-shared

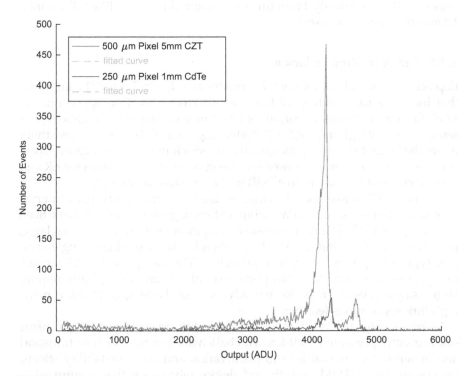

FIGURE 2.14
Comparison between 5 mm thick CZT sensor (red) and 1 mm thick CdTe sensor (blue) exposed to Co-57 radiation. Courtesy: Rutherford Appleton Laboratory.

events. Excellent ER of the CZT sensor (better than 2 keV) is clearly visible due to clean separation between two Co-57 peaks at 122 keV and 136 keV.

In order to prepare for high-volume commercial production, the CZT industry is moving from individual tile processing to whole-wafer processing using silicon-like methodologies. Parametric-level screening is being developed at the wafer stage to ensure high wafer quality before detector fabrication in order to maximize production yields. These process improvements enable CZT manufacturers to provide high-volume production for photon-counting applications in an economically feasible manner.

CZT sensors are capable of delivering both high-count rates and high-resolution spectroscopic performance although it is challenging to achieve both of these attributes simultaneously. It is possible to achieve 1 keV ER in spectroscopy at 60 keV, which is not as good as HPGe but better than GaAs, and 300 Mcps/mm² flux rate in photon counting, significantly better than CdTe. In fact, CZT sensors provide the highest flux rate response from high-Z materials, therefore, several large medical imaging OEMs are building next generation CT scanners using this technology. Recent publications discuss material challenges, detector design trade-offs and hybrid pixel readout chip architectures required to build cost-effective CZT-based detection systems [16–21].

2.3.6 Comparison between Semiconductor Materials

There has been some effort in the research community to produce new semiconductor materials that can compete with CZT, such as Thallium Bromide (TiBr$_4$) and Mercury Iodide (HgI$_2$). A basic summary of semiconductor material properties is shown in Table 2.1. At present, none of these materials are nowhere near production readiness required for volume manufacturing so they are not discussed any further here.

Each sensor technology presents its own limitation on the energy range of incoming photons that can be detected. The limitation on the low end side typically comes from the readout ASIC electronics not from sensor properties. A more detailed discussion on this topic and description of the photon-counting ASIC electronics is provided later in Chapter 3 in this book so, at this point, we will only make some estimates on the minimum photon energy (E_{min}) that can be detected.

TABLE 2.1

Basic Properties of Selected Semiconductor Materials Used for Radiation Detection

Material	Ge	Si	CZT	TlBr	HgI2
Average Z	32	14	49	80	80
Density (g/cm³)	5.32	2.33	5.78	7.56	6.4
Resistivity (Ohm-cm)	50	1E5	1E10	1E12	1E12

A useful rule of thumb is that E_{min} is 3× ER of the system. This condition ensures that there are no false trigger events that are detected due to noise events. The energy resolution ER is a function of both sensor and the ASIC electronics. Typical values for high-Z materials are given in Table 2.2. As shown there are no relevant differences between all sensor materials (with the exception of GaAs) unless very low energies are of interests.

The maximum photon energy is in practice related to the maximum thickness of the sensor. If the sensor is not thick enough, a significant portion of the incoming radiation will escape the sensor material. This has severe undesired consequences. First, sensor efficiency suffers leading to required long exposure time in order to collect sufficient number of photons to ensure good signal-to-noise (SNR) and contrast-to-noise (CNR) values. Second, escaping radiation will cause radiation damage in the underlying electronics. The degree of the radiation damage depends on the radiation hardness of the electronics, but it will lead to high implementation costs (if rad hard technologies are used) and/or high servicing costs (if frequent replacement costs have to be accepted).

As discussed earlier, Si, Ge and GaAs are limited to standard 500 μm thickness typical for microelectronics industry. As a result, to obtain 90% DQE their maximum photon energy is severely limited as shown in Table 2.2 being as little as 12 keV for silicon and only 35 keV for Ge and GaAs. CdTe sensors can be thicker, but the polarization effect limits in practice their thickness to 1 mm which corresponds to the maximum energy of 60 keV. CZT sensors do not exhibit polarization effects, if properly designed and fabricated, therefore, even 5 mm thick material might be suitable for NDT applications extending the energy range significantly to 1,000 keV. CT scanners that require very high-count rates typically use 2–3 mm thick CZT.

The values calculated in Table 2.2 are for 90% absorption level in the sensor. If lower absorption levels are acceptable, the energy range can be obviously extended. CZT sensors are used in CT scanners up to 160 keV and can easily operate at high count rate for energies up to 200 keV. This range of photon energies is more than sufficient for XRD applications.

TABLE 2.2

Typical Values for Energy Resolution (ER), Minimum Photon Energy E_{min}, Maximum Sensor Thickness and Maximum Photon Energy E_{max} for Selection of the High-Z Sensors

Material	Ge	Si	GaAs	CdTe	CZT
Energy resolution (keV)	0.5	0.5	3	1	1
Minimum energy (keV)	1.5	1.5	9	3	3
Maximum thickness (mm)	0.5	0.5	0.5	1	5
Maximum energy (keV)	35	12	35	60	100

2.4 Sensor Technology Examples

To illustrate how direct conversion sensor technology can be deployed in XRD applications, we describe in this section some examples of development work with various integrated circuits (ICs.) As described above, high-count rate operation of CZT can be only effectively verified using integration with pixelated ASICs. Some of the experimental results are discussed in the following sections. Design and operation of these ASICs is described in detail in Chapter 3.

2.4.1 ASIC Electronics

Semiconductor pixelated detectors need to have a high level of segmented multi-channel readout. Several decades ago, the only way to achieve this was via massive fan-out schemes to route signals to discrete low-density electronics. At the present time, CMOS technology is used to build very dense low-power electronics with many channels which can be bonded directly or indirectly (through a common carrier printed circuit board PCB) to the detector.

There are different requirements for the CMOS technology used for the analog front-end signal processing, as opposed to that of the digital signal processing. For the analog part of the electronics, there is a requirement for a robust technology that has low electronic noise and high dynamic range that typically requires high-power supply voltages. Digital signal processing in turn requires very high speed and high density that is more compatible with the more modern low voltage supply, deep submicron processes. Design and operation of readout ASIC electronics is a complex issue, as a result, the entire Chapter 3 is devoted to this topic.

2.4.2 ASIC to Sensor Attachment

CZT pixel detectors require a connection from the pads on the detector material to the bond pads on the ASICs. In some cases, the pixel pitch on the readout ASIC is the same as the detector pixel pitch. It is also possible to fan out the connections on the detector with multilevel metal routing on the detector or with the use of the interposer board. This fanning out routing has to be done carefully as there is a large danger for signal crosstalk. With integrating readout and synchronous input signals, where the signal is totally removed from the detector, this might not be a problem but transients can still upset thresholds in these systems.

The pixel pitch of X-ray imaging systems currently falls between about 100 μm and 1 mm. For small pixel pitches, bump bonding is used to connect the detector pixels to the ASICs. There are many different technologies to do this depending on the requirements of the detectors and environmental constraints. The industry standard area bump-bonding method is to deposit solder onto under-bump metallization on the pads of the detector and ASIC and then to align the two and heat them to reflow the solder, as schematically shown in Figure 2.15. Various solders are used including lead-tin, bismuth-tin, indium alloys, silver alloys depending on the temperature to reflow and the operating temperature required. Typically, these materials require 240°C– 140°C to reflow. Indium is used either in a lower temperature reflow process or straight compression bonding and gives good results but cannot be used if high-operating temperatures will ever be experienced.

With all these processes, either fluxes or special gas environments including nitrogen, hydrogen or formic acid have to be used to ensure good contact between parts. These processes can allow 10–20 μm bond diameters to be used which allow 50–100 μm pixels to be connected. Another method used particularly for larger pixel (above 150 μm) CZT detectors relies on gold studs and silver-loaded epoxy dots. The gold studs are either bonded or deposited on the ASIC part and the silver-loaded epoxy is screen printed on the detector.

2.4.3 Sensors Performance Examples

This section describes some examples of direct conversion detector performance characteristics using several different ASIC Integrated Circuits for both PC and spectroscopy modes. The ASIC operation is described in Chapter 3.

2.4.3.1 Pilatus Photon-Counting Readout

1.5 mm thick CZT detectors with Platinum (Pt) contacts have been fabricated at Redlen Technologies and directly attached to PILATUS3 ASIC with

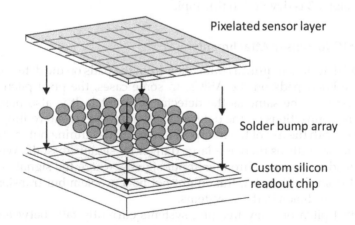

Pixelated sensor layer

Solder bump array

Custom silicon readout chip

FIGURE 2.15
Schematic illustration of the bump-bonding process that connects the sensor and ASIC readout chip.

re-triggering capability. The resulting 172 µm pixel pitch module was measured by the Dectris group using the same characterization procedure as used for CdTe and Si sensors. The samples were characterized using the following characterization steps: optical inspection, flat-field measurements (using 25 keV fluorescence source), direct beam tests (tungsten X-ray tube with 60 kVp and 3 mm Al filter), followed by bias voltage, where full charge collection is achieved and leakage current measurements [21].

In order to test sensor behavior with increasing count rate, the X-ray tube current was increased in steps from 1 mA up to 19 mA as shown in Figure 2.16. CZT sensors operated successfully up to the limits of the measurement set-up of 6.2 Mcps/channel (which translates to 209 Mcps/mm^2 considering the 172 µm pitch). The measured output count rate was 5.5 Mcps/mm^2 which indicates 12% pile-up loss at that rate. No polarization effects in the sensor were observed. Measured count rate stability was about 0.3% within 1 hour measurement window.

2.4.3.2 DxRay Photon-Counting Readout

2 mm thick CZT detectors with Platinum (Pt) contacts have been fabricated at Redlen Technologies and temporarily attached using pogo-pin set-up to

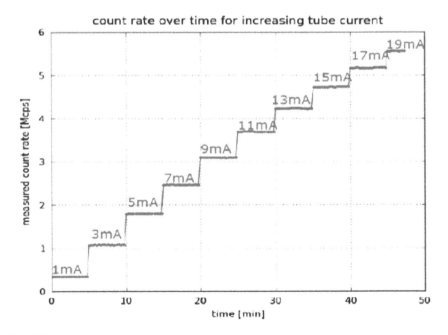

FIGURE 2.16
Measured count rate (Mcps) as a function of the X-ray tube current. A polychromatic X-ray field from a tungsten X-ray tube with 60 kVp and 3 mm Al filter. The detector threshold is set to 8 keV [21].

1 mm pixel pitch DxRay photon-counting ASIC [21]. Charge collection tests using 60 keV X-ray tube with 0.5 mm copper (Cu) filter indicate that high voltage (HV) of 400 V was sufficient to collect the generated charges as any higher voltage (tested up to 600 V) did not improve detector efficiency. The measured spectra, shown in Figure 2.17, did not vary with the count rate within range used (from 1 Mcps to 4 Mcps/channel).

2.4.3.3 Timepix Photon-Counting Readout

2 mm thick CZT sensors with 110 μm pixels were bonded to Timepix ASIC. The resulting CZT Timepix assembly mounted to the ModuPIX readout [21]. The bias voltage range for MuduPIX was −500 V and maximal frame-rate: 850 fps (USB 2.0) or 1700 fps (USB 3.0). During the set-up, detector was equalized to assure uniform threshold for comparators of all pixels (performed with moderate bias of −300 V). The threshold level was calibrated performing threshold scan with gamma line of Am-241 and XRF K lines of Cd and Cu. 100% bump-bonding yield (all pixels connected, no cross connections).

The leakage current was successfully compensated increasing the Krummenacher current in Timepix preamplifiers (I_{krum}). The final I_{krum} value typically used was 5 (indicating leakage compensation of about 0.4 nA) which is of moderate value. Other High-Z sensors require usually

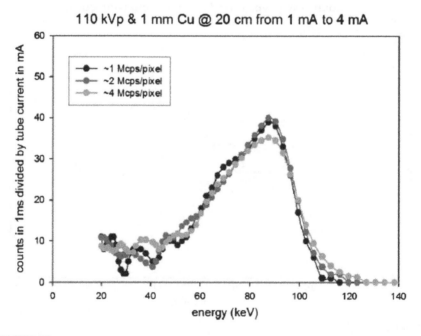

FIGURE 2.17
Measured X-ray spectrum with 110 keV X-ray tube setting with 1 mm thick copper filter at 20 cm distance at 3 values of the count rate [21].

higher values; for example, CdTe sensors usually require about twice as much, i.e., I_{krum} of at least 10, sometimes as high as 40. In all tests performed, minimal threshold of 3.5 keV was set. Results for NDT applications are reported elsewhere [28–38].

2.4.3.4 HEXITEC Spectroscopic Readout

It has been shown very recently that traditional high-energy X-ray systems can capture scattered X-rays to deliver 3D images with structural or chemical information in each voxel [39]. This type of imaging can be used to separate and identify chemical species in bulk objects with no special sample preparation. The capability of hyperspectral imaging has been demonstrated by examining an electronic device where the atomic composition of the circuit board components in both fluorescence and transmission geometries can be easily identified.

Hyperspectral images were recorded using the HEXITEC camera, operating under an applied bias voltage of 500 V at a frame rate of 20 kHz and a maximum count rate of about 5 kcps/mm². The HEXITEC camera consists of a 1 mm thick CdTe single crystal detector (20 × 20 mm²) bump-bonded to a large area ASIC packaged with a high performance data acquisition system.

The detector has 80 × 80 pixels of size 250 mm with an energy resolution (FWHM) of about 800 eV @ 59.5 keV and about 1.5 keV @ 141 keV. During operation, each photon event has its charge and pixel position, and the frame in which it occurs is recorded. This strategy supports measurements when photon timing is relevant though it was not exploited in this case. Events are processed and histograms created according to measured charge, using a total of 400 bins. During this process, a charge-sharing strategy is employed to deal with events that may have resulted from the sharing of charge between two or more pixels (these being a subset of neighboring events in a single frame). Charge discrimination was employed to perform fluorescence imaging. The images were energy calibrated such that each pixel had a common energy axis; the calibrations for each pixel were found from their response to the fluorescence from a set of pure metals. More details of this readout technology can be found in reference [39].

To demonstrate the possibilities of hyperspectral methods, a USB wireless web-cam dongle was radiographed using both a traditional and hyperspectral X-ray camera using a standard laboratory tungsten target X-ray tube operating at 80 kV. Since a hyperspectral image contains an energy spectrum for every pixel, it is possible to select images at different energies or to extract spectra from different regions of the image. By exploiting the extra information contained in the spectral domain, the researchers were able to gain additional insight into the sample using multivariate analysis methods to find natural groupings in the data based on their spectral similarity. Applying principal component analysis to the data, they were able to separate out certain features. Plastic casing, printed circuit board and higher density metal features were clearly identified based upon their spectral profiles alone.

2.5 Conclusions

The purpose of this chapter was to describe semiconductor sensors used for direct X-ray detection. In the first section, we have described typical X-ray detection applications and resulting sensor requirements. In the second section, we have contrasted traditional scintillator detectors with modern direct conversion detector technology using semiconductor detectors. In the third section, we have analyzed direct conversion sensors and contrast Si, Ge, GaAs, CdTe and CZT. We have showed why semiconductor sensors are typically the best choice, in particular for high-count and/or high-energy applications. In the fourth section, we have discussed requirements of the sensor electronics required to read signals generated by the direct conversion sensors and demonstrated some example of usage of semiconductor sensors in NDT, medical imaging including CT and baggage scanning. Most of these examples are directly applicable to X-ray diffraction technologies.

Based on the review of available sensor technologies mentioned above, CZT seems like the most suitable choice for high-count and/or high-energy applications. Recent advances in THM growth and device fabrication have dramatically improved hole transport properties and reduced the polarization effects in CZT material. As a result, high flux operation of CZT sensors at rates in excess of 200 Mcps/mm^2 is now possible and has enabled multiple medical imaging companies to start building prototype CT scanners and high-efficiency Linear Array Detectors. Due to economy of scale, CZT sensors are now finding new commercial applications in NDT and baggage scanning. CZT sensors are capable of delivering both high-count rates and high-resolution spectroscopic performance although it is challenging to achieve both of these attributes simultaneously. For X-ray diffraction applications, the key requirement is good ER. In this respect, CZT detectors can achieve ER down to the 1 keV level if a suitable ASIC architecture is selected, see [22–27] for examples of available ASIC chips.

References

1. *Source*: www.i3system.com/eng/n_tech/tech2.html.
2. B. J. Heismann, D. Henseler, D. Niederloehner, P. Hackenschmied, M. Strassburg, S. Janssen, and S. Wirth, Spectral and Spatial Resolution of Semiconductor Detectors in Medical X- and Gamma Ray Imaging, in *Medical Imaging Technology and Applications*, K. Iniewski and T. Farncombe (Eds.), Medical Imaging, CRC Press, Boca Raton (2013). eBook ISBN 9781466582637.

3. K. Zuber et al., *A Comparison of the Temporal Instabilities Found in A Comparison of the Temporal Instabilities Found in State-of-the-Art CdTe and CZT Sensors Used in Spectral CT Measurements with the Medipix3RX Detector*, IEEE NSS-MICS 2015, San Diego.

4. R. Redus, *Charge Trapping in XR-100-CdTe and -CZT Detectors*, Amptek, 2007.

5. R. Macdonald, Design and Implementation of a Dual-Energy X-Ray Imaging System for Organic Material Detection in an Airport Security Application, *Proc. of SPIE 4301* (2001) 65–69.

6. I. Farella, Study on Instability Phenomena in CdTe Diode-Like Detectors, *IEEE Trans. Nucl. Sci.* 56(4) (September 2009) 1736–1742.

7. H. Toyama et al., *Analysis of Polarization Phenomenon and Deep Acceptor in CdTe Radiation Detectors*, IEEE NSS-MICS, 2006.

8. A. Arodzero et al., *A System for the Characterization and Testing of CdZnTe/ CdTePixel Detectors for X-Ray and Gamma-ray Imaging*, IEEE NSS-MICS, 2006.

9. T. Seino, S. Kominami, Y. Ueno, and K. Amemiya, Pulsed Bias Voltage Shutdown to Suppress the Polarization Effect for a CdTe Radiation Detector, *IEEE Trans. Nucl. Sci.* 55(5) (October 2008) 2770–2774.

10. L. Verger, E. Gros d'Aillon, O. Monnet, G. Montemont, and B. Pelliciari, New Trends in γ-Ray Imaging with CdZnTe/CdTe at CEA-Leti, *Nucl. Instrum. Meth. A* 571 (February 2007) 33–43.

11. X. Wang, D. Meier, B. Sundal, B. Oya, P. Maehlum, G. Wagenaar, D. Bradley, E. Patt, B. Tsui, and E. Frey, A Digital Line-Camera for Energy Resolved X-Ray Photon Counting, *Proc. IEEE Nucl. Sci. Symp. Rec.*, 2009, pp. 3453–3457.

12. A. Brambilla, C. Boudou, P. Ouvrier-Buffet, F. Mougel, G. Gonon, J. Rinkel, and L. Verger, Spectrometric Performances of CdTe and CdZnTe Semiconductor Detector Arrays at High X-Ray Flux, *Proc. IEEE Nucl. Sci. Symp. Rec.*, 2009, pp. 1753–1757.

13. J. Rinkel, G. Beldjoudi, V. Rebuffel, C. Boudou, P. Ouvrier-Buffet, G. Gonon, L. Verger, and A. Brambilla, Experimental Evaluation of Material Identification Methods with CdTe X-Ray Spectrometric Detector, *IEEE Trans. Nucl. Sci.* 58(2) (October 2011) 2371–2377.

14. L. Tlustos, Spectroscopic X-Ray Imaging with Photon Counting Pixel Detectors, *Nucl. Instrum. Meth. A* 623(2) (November 2010) 823–828.

15. J. Iwanczyk, E. Nygard, O. Meirav, J. Arenson, W. Barber, N. Hartsiugh, N. Malakhov, and J. Wessel, Photon Counting Energy Dispersive Detector Arrays for X-Ray Imaging, *Proc. IEEE Nucl. Sci. Symp. Rec.*, 2007, pp. 2741–2748.

16. C. Szeles, S. Soldner, S. Vydrin, J. Graves, and D. Bale, CdZnTe Semiconductor Detectors for Spectrometric X-Ray Imaging, *IEEE Trans. Nucl. Sci.* 55(1) (February 2008) 572–582.

17. E. Kraft and I. Peric, Circuits for Digital X-Ray Imaging: Counting and Integration, in *Medical Imaging Electronics*, K. Iniewski (Ed.), Wiley, New York (2008).

18. S. Mikkelsen, D. Meier, G. Maehlum, P. Oya, B. Sundal, and J. Talebi, An ASIC for Multi-energy X-Ray Counting, *Proc. IEEE Nucl. Sci. Symp. Rec.*, 2008, pp. v294–299.

19. S. Awadalla, Solid-State Radiation Detectors: Technology and Applications, in *Devices, Circuits, and Systems*, CRC Press, Boca Raton, FL (2015).

20. G. Prekas et al., The Effect of Crystal Quality on the Behavior of Semi-Insulating CdZnTe Detectors for X-Ray Spectroscopic and High Flux Applications, *2014 IEEE Nuclear Science Symposium and Medical Imaging Conference*, Seattle, WA, U.S.A., November 8–15 2014.

21. K. Iniewski, CZT Growth, Characterization, Fabrication and Electronics for operation at 100 Mcps/mm^2, *Workshop on Medical Applications of Spectroscopic X-Ray Detectors, CERN,* Geneva, Switzerland, April 20–23 2015.
22. J. Iwanczyk, Radiation Detectors for Medical Imaging, in *Devices, Circuits, and Systems,* CRC Press, Boca Raton Florida U.S.A. (2015).
23. R. Turchetta, Analog Electronics for Radiation Detection, in *Devices, Circuits, and Systems,* CRC Press, Boca Raton Florida U.S.A. (2016).
24. R. Steadman et al., ChromAIX, A High-Rate Energy Resolving Photon-Counting ASIC for Spectral Computed Tomography, *Proc. SPIE 7622* (2010) 762220.
25. W. Barber et al., Photon Counting Energy Resolving CdTe Detectors for High-flux X-Ray Imaging, *2010 IEEE Nuclear Science Symposium and Medical Imaging Conference,* Knoxville, TN, October 30–November 6 2010, p. 3953.
26. C. Ullberg et al., Measurements of a Dual-Energy Fast Photon Counting CdTe Detector with Integrated Charge Sharing Correction, *Proc. SPIE 8668* (2013) 86680P.
27. R. Ballabriga et al., *The Medipix3RX: A High Resolution, Zero Dead-Time Pixel Detector Readout Chip Allowing Spectroscopic Imaging,* 2013 JINST 8 C02016.
28. P. Maj et al., Measurements of Matching and Noise Performance of a Prototype Readout Chip in 40 nm CMOS process for Hybrid Pixel Detectors, *IEEE Trans. Nucl. Sci.* 62 (2015) 359.
29. G. Deptuch et al., Results of Tests of Three-Dimensionally Integrated Chips Bonded to Sensors, *IEEE Trans. Nucl. Sci.* 62 (2015) 349.
30. R. Ballabriga et al., *Review of Hybrid Pixel Detector Readout ASICs for Spectroscopic X-Ray Imaging,* 2016 JINST 11 P01007.
31. P. Kraft et al., Characterization and Calibration of PILATUS Detectors, *IEEE Trans. Nucl. Sci.* 56 (2009) 758.
32. T. Koenig et al., How Spectroscopic X-Ray Imaging Benefits from Inter-Pixel Communication, *Phys. Med. Biol.* 59 (2014) 6195.
33. K. Iniewski et al., High Precision Medium Flux Rate CZT Spectroscopy, *2015 IEEE Nuclear Science Symposium and Medical Imaging Conference,* San Diego, CA, October 31–November 7 2015.
34. K. Taguchi and J. S. Iwanczyk, Vision 20-20: Single Photon Counting X-Ray Detectors in Medical Imaging, *Med. Phys.* 40 (2013) 100901.
35. S. Kappler et al., A Research Prototype System for Quantum-Counting Clinical CT, *Proc. SPIE 7622* (2010) 76221Z.
36. A. Altman, *State of the Art and Future Trends in Radiation Detection for Computed Tomography,* 2013, AAPM Virtual Library, www.aapm.org/education/VL/vl.asp?id=2325.
37. K. Mahnken et al., Spectral RhoZ-Projection Method for Characterization of Body Fluids in Computed Tomography: Ex-Vivo Experiments, *Acad. Radiol.* 16 (2009) 763–769.
38. K. Iniewski and J. Jakubek, *Small Pixel CZT Characterization for Non-Destructive Testing,* in preparation.
39. S. D. M. Jacques, C. K. Egan, M. D. Wilson, M. C. Veale, P. Seller, R. J. Cernik, A Laboratory System for Element Specific Hyperspectral X-Ray Imaging, *Analyst* 138 (2013) 755–759, The Royal Society of Chemistry 2013.

3

Integrated Circuits for XRD Imaging

Krzysztof Iniewski
Redlen Technologies

Chris Siu
British Columbia Institute of Technology (BCIT)

CONTENTS

3.1 Introduction

As discussed in Chapter 2, XRD application requires specialized circuit electronics. Early developments in this field were using discrete electronics, but over the past decade, the signal processing has been done mostly with Integrated Circuits (ICs), also called Application-Specific Integrated Circuits (ASICs), that are either directly or through the joining interposer attached to the semiconductor sensors. Modern integrated circuit technology makes it economically feasible for each pixel to have its own low-noise electronics, creating what are known as "hybrid pixel detector readout ASICs." The purpose of this chapter is to review some fundamentals of this readout IC electronics and present some examples of its implementation by various industrial vendors and research organizations worldwide.

Radiation detectors detect and convert radiation into electric signals. The amount of charge, Q, generated by an ionizing event is proportional to energy deposited in the detector. Almost all X-ray detection applications require either discriminating or measuring Q and, in some cases, its time of arrival. The detector has at least two electrodes but frequently has a larger number, especially on one of its sides (e.g., pixelated-, strip-, co-planar, or Frish grid-detectors, as discussed in Chapter 2). An electric field is generated in the detector, most often by applying a voltage between the pixelated (or segmented) electrodes on one side and a common electrode on the opposite side. Under this electric field, the ionized charge, Q, moves toward one or more electrodes inducing a charge flow in each.

Depending upon the type of detector and the particular application, the charge can be read out event by event or it can be integrated from several events in the capacitance of the electrode and read out at a later time. In most cases discussed in this book, photons arrive randomly in time domain. Invariably, reading out the signals from radiation detectors requires highly specialized electronics, usually referred to as "front-end" electronics. This electronics usually entails stringent requirements in terms of the signal-to-noise ratio, dynamic range, linearity, and stability.

A typical front-end channel is composed of three fundamental blocks: the low-noise amplifier, the filter (usually referred to as the "shaper"), and the peak detector. The low-noise amplifier reduces the relative noise contribution from the next blocks, i.e., the shaper and peak detector, to negligible levels. The shaper is required to optimize the signal-to-noise ratio, while the peak detector (PD) provides discrimination, measurement, and storage depending on the application (e.g., waveform discrimination and counting, waveform peak and/or timing, and periodic sampling).

Application of semiconductor detectors requires use of dedicated detector electronics. The signal generated by X-ray sensors needs to be amplified, filtered, and possibly stored by analog circuitry for subsequence conversion to digital signal domain for further processing. Analog signal processing requires use of dedicated discrete electronics or application-specific integrated circuit (ASIC). Digital processing typically requires use of FPGA to control analog front end.

3.2 Readout ICs Fundamentals

3.2.1 IC Technology

Semiconductor-pixelated detectors need to have a high level of segmented multi-channel readout. Several decades ago, the only way to achieve this was via massive fan-out schemes to route signals to discrete low-density electronics. At the present time, CMOS technology is used to build very dense low-power electronics with many channels which can be bonded directly or indirectly (through a common carrier PCB) to the detector.

There are different requirements for the CMOS technology used for the analog front-end signal processing, as opposed to that of the digital signal processing. For the analog part of the electronics, there is a requirement for a robust technology that has low electronic noise and high-dynamic range that typically requires high power supply voltages. Digital signal processing in turn requires very high speed and high density that is more compatible with the more modern low-voltage supply, deep submicron processes.

There seems to be a technology optimum at around 0.35 µm to 0.18 µm minimum feature size for the analog requirements. The large feature size limits the complexity of circuitry that can be integrated in a pixel but even at 0.35 µm it is possible to place a million transistors on a reasonable size silicon die. In comparison, digital signal processing can benefit from the rapid development of deep submicron processes. Some selected research developments now take place using 90 nm or 65 nm process nodes. These technologies are well suited to high-speed ADC architectures and to very fast data manipulation for data sparcification and compression. The deep submicron technologies have their own limitations in terms of analog performance, noise, and cost.

A block diagram of the typical charge detection electronics for each pixel or channel is shown in Figure 3.1 [1–3].

The charge generated by the semiconductor sensor Q_0 (in this example, CZT sensor is being shown) enters a Charge-Sensitive Amplifier (CSA), which converts the charge to a voltage using the feedback capacitor C_f:

$$V_{CSA} = \frac{Q_0}{C_f}$$

As an example, let us assume C_f value of 10 fF. Since a 59.5 eV photon produces 1.9 fC of charge, this corresponds to a step change of 190 mV at the CSA output, a reasonably large value to be detected.

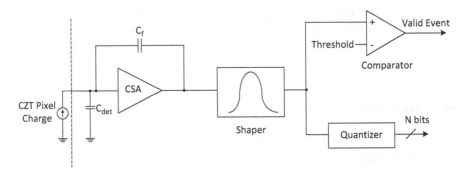

FIGURE 3.1
Charge detection electronics block diagram.

The transient V_{CSA} step then enters a bandpass filter to improve the signal-to-noise ratio. Since this bandpass filter changes the shape of the V_{CSA} response, it is also called a pulse shaper. The peak output of the shaper is compared to a programmable threshold to determine if a valid pixel event has occurred; this comparison is done to remove false events due to noise. Finally, a valid pixel event is digitalized using an ADC for further downstream processing. An example of signal processing is shown in Figure 3.2.

Front-end ASIC electronics is an important part of the entire signal processing chain. However, in order to facilitate the integration of semiconductor sensors in end-customer XRD systems, a further signal processing is required, frequently referred to as X-ray detection module. A block diagram of such a module is shown in Figure 3.3. Ethernet PHY block represents I/O

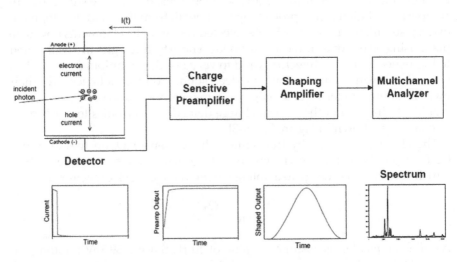

FIGURE 3.2
Typical output from each stage of the detector electronics [4].

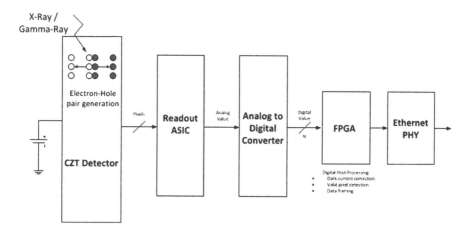

FIGURE 3.3
Complete readout signal processing block diagram.

transceiver as an example, as other I/O technologies like Camera Link, Fibre Channel, or Rapid I/O Phy can be used as well depending on the requirement on the connecting external PC host environment.

3.2.2 IC Design Process

IC design process contains the following phases to take place until device tape-out (TO) to the selected foundry as schematically illustrated in Figure 3.4.

- Feasibility Phase – During the feasibility phase, the design team is expected to perform architectural development for the proposed readout IC solution. All relevant process and fabrication arrangements have to be finalized at that stage by the design team. Computer Aided Design (CAD) technology files have to be installed in the IC design flow at that time. The feasibility phase ends with a Design Start review where the design team discusses with the proposed solution and all outstanding issues, if any, are clarified.

- Design/Simulation Phase – During the simulation phase, the design team is expected to perform SPICE simulation results for the detailed circuit implementation of the IC.

- The design phase ends with a Design Simulation review where the design team presents detailed circuit simulation results. It is expected that the design team will provide a written engineering report that contains circuit simulation results before proceeding to the next stage as required by concurrent design and documentation good practices.

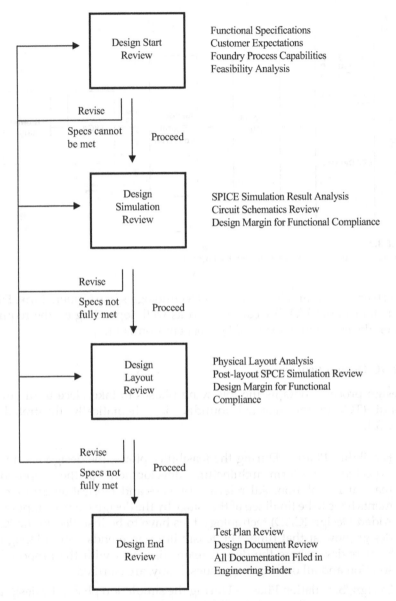

FIGURE 3.4
IC design process sequence.

- Layout Phase – During the layout phase, the design team is expected to perform IC layout in a given manufacturing process. All layout implementation issues have to be finalized by the design team at that stage. Any post-layout simulation results that would indicate a necessity to revise circuit schematics have to be reported immediately to the designated representative. The layout phase ends with a Design Layout review. It is expected that the design team will provide an updated written engineering report that contains post-layout circuit simulation results before proceeding to the next stage (concurrent design and documentation).

- Verification Phase – During verification phase, the design team is expected to perform exhaustive verification of the completed design before submitting the design for tape-out (TO). For example, device mismatch, power supply noise, substrate crosstalk, I/O ground bounce, etc. have to be taken into account. ASIC verification for reliability requirements: interconnect/via electromigration, ESD, and latch-up is expected as well. It is essential that the verification be performed in conjunction with the CZT detector model, package, and PCB parasitic elements. During that phase, the design team completes a test plan for the prototype evaluation phase to be followed. It is expected that the entire engineering documentation is complete before proceeding to the tape-out (TO). The verification phase ends with Design End review.

3.2.3 Photon Counting vs. Spectroscopy

There are two ways of signal processing for XRD with energy-sensitive semiconductor detectors: photon counting and spectroscopy. While precise difference between the two might be hard to establish as all analog signals eventually become digital at some point in the readout system, we would like to suggest the following practical definition. Photon counting relies on energy binning using comparators inside the readout IC chip. Spectroscopy in turn preserves the analog nature of the signal representing photon energy with an A/D converter after the readout IC signal processing has been accomplished.

As a result of this architectural change, photon counting systems can achieve very high count rate while sacrificing energy resolution (ER), while spectroscopic system can have very good noise properties but face limitations with a maximum count rate. To achieve specific design objectives, either count rate of the spectroscopic system or energy resolution of the photon-counting systems is maximized. Schematic comparison between two types of signal processing is shown in Figure 3.5.

3.2.4 Photon Counting ICs

One of the major advantages of photon-counting detectors is electronics noise rejection. Well-designed photon-counting detectors allow for ASIC electronics threshold high enough to reject noise pulses while still counting useful signals. Therefore, quantum limited operation of the photon-counting detector can be achieved as image noise is determined only by statistical

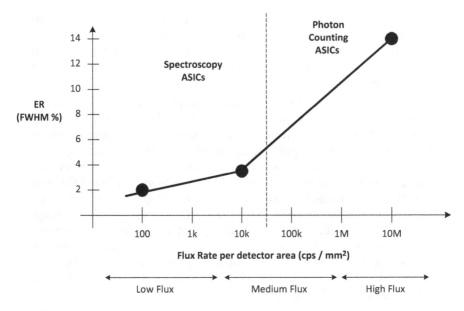

FIGURE 3.5
Schematic comparison between low flux spectroscopic ICs and high-flux photon-counting ICs.
Energy resolution (ER) is represented at 60 keV level.

variations of X-ray photons. On the other hand, energy-integrating detectors
suffer from electronics noise which is mixed with useful photon signals and
separating it from statistical noise is not possible. Electronics noise rejection
is important because its magnitude for currently used digital X-ray detectors
is not negligible.

After converting the CZT-generated charge to voltage by the CSA and
subsequent filtering by the shaping amplifier, the signal is ready for digitization. Typically, the signal is compared against user selected threshold
voltage (discriminator box in Figure 3.6) to produce 1-bit trigger signal indicating detection of the pulse. In parallel, the value of the shaped signal is
sent to an ADC converter (or Time over Threshold, or ToT, processor) with
n-bit accuracy. The conversion resolution n is typically between 8 and 16
bits depending on the system accuracy, noise levels, and degree of signal
precision achieved. One important consideration in the practical system is
CSA reset. As the feedback capacitor C_f is charged by the input signal, there
must be some means of discharging this capacitor in order for the CSA to be
ready for the next signal. This circuitry is schematically shown as the reset
block in Figure 3.6.

There are two possible implementations for the reset block: digital and
analog. The digital one involves using a switch that will discharge the feedback capacitor quickly. Unfortunately, this process typically creates too
much disturbance for the sensitive CSA. The analog solution involves using

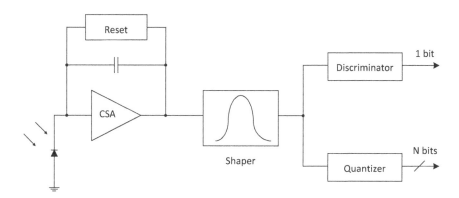

FIGURE 3.6
Photon-counting detector readout signal chain.

a resistor (or MOSFET operating in the triode region) and provides continuous discharging during the entire process. The discharge cannot be too slow (in which case, the capacitor will not be fully discharged before the next event) or too fast (as that will affect signal formation). While the CZT readout scheme shown in Figure 3.6 is a typical implementation, it is possible to directly sample the signals without producing any trigger signal to obtain timing and amplitude information.

On a final note, while the principle of CSA signal amplification, pulse shaping, and ADC conversions outlined above are fairly simple, practical implementations can be very challenging due to very small input signals involved (below 1 mV). One has to pay particular attention to system noise, power supply decoupling, ESD protection, EMI radiation, and op-amp stability issues.

A typical photon counting ASIC implementation contains hundreds of channels frequently implemented with multiple energy bins. A typical 128-channel ASIC architecture is shown in Figure 3.7.

3.2.5 Spectroscopic ICs

XRD detectors typically operate in a single photon detection mode where an electric charge generated by one photon needs to be collected by the readout electronics. As the amount of generated charge is small (few femto Columbs, fCs), very sensitive analog circuitry is required to amplify that charge. In spectroscopic applications, the amount of charge, which directly corresponds to the photon energy, needs to be precisely determined. In photon-counting applications, only binary (or multi-binary) decision is required but the count rate might be very high creating its related challenges. The purpose of this section is to explain some of the design considerations that are important when building XRD readout electronics systems.

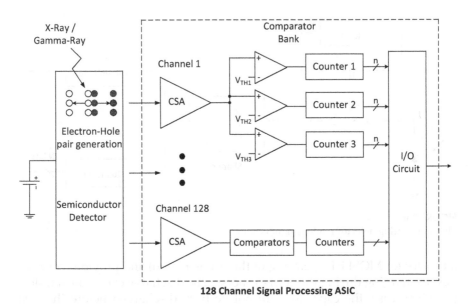

FIGURE 3.7
Block diagram of 128-channel photon-counting readout IC.

3.2.5.1 Analog Front-End

Analog signal processing can be divided into the following steps:

- Amplification – The input charge signal is amplified and converted to a voltage signal using a CSA. The main characteristic of the amplification stage is Equivalent Noise Charge (ENC) which is required to be as low as possible in order not to degrade intrinsic detector energy resolution. Another important consideration for the CSA operation is a dark current compensation mechanism. A solution that accommodates continuous compensation for dark currents up to several nAs while maintaining low ENC is desired.

- Signal shaping – The time response of the system is tailored to optimize the measurement of signal magnitude or time and the rate of signal detection. The output of the signal chain is a pulse whose area is proportional to the original signal charge, i.e. the energy deposited in the detector. The pulse shaper transforms a narrow detector current pulse to broader pulse (to reduce electronic noise), with a gradually rounded maximum at the peaking time to facilitate measurement of the amplitude. A solution that provides effective signal shaping while maximizing the channel count rate needs to be applied.

- Pulse detection – The input pulse, broadened by the shaping process, needs to be detected against a set-up threshold value. The threshold level is a critical parameter that determines whether the event is recognized as a true event or false reading caused by noise. As a result, the threshold value is typically adjustable both globally and at the pixel level. The peak detection value determines energy-level information. A solution that prevents temperature drift of the PD needs to be used.

- Channel multiplexing – In case of ASIC spectroscopy, all parallel channels of the channel readout ASIC need to have their signals multiplexed at the output before being sent out to an external ADC. The key requirement to channel multiplexing and signal shaping is a maximum channel count rate determined by the given application.

3.2.5.2 Charge-Sensitive Amplifier

For single photon detection, very sensitive analog circuitry is required to amplify the charge generated by the X-ray sensor. The current signal induced in the sensing electrode can be integrated in the pixel capacitance and read out with a high input impedance stage, which amplifies the resulting voltage at the pixel node or it can be read out directly with a low input impedance stage, which amplifies the charge Q and keeps the pixel node at a virtual ground, such as a charge amplifier. The latter is the preferred choice since, among its other advantages, it stabilizes the sensing electrode by keeping its voltage constant during the measurement and/or the read out.

In both cases, low-noise amplification is required to reduce the noise contribution from the processing electronics (such as the shaper, peak detector, and ADC) to negligible amount; good design practice dictates maximizing this amplification while avoiding overload of subsequent stages. This low-noise amplification would also provide either a charge-to-voltage conversion (e.g., source follower, charge amplifier) or a direct charge-to-charge (or current-to-current) amplification (e.g., charge amplifier with compensation, current amplifier). Depending upon this choice, the shaper would be designed to accept a voltage or a current, respectively, as its input signal.

In a properly designed low-noise amplifier, the noise is dominated by processes in the input transistor. Assuming that CMOS technology is employed in the design, the input transistor is referred to as the "input MOSFET" although the design techniques can easily be extended to other types of transistors, such as the JFET, the bipolar transistor, or the hetero-junction transistor. The design phase, which consists of sizing the input MOSFET for maximum resolution, is called "input MOSFET optimization" and has been studied extensively in the literature.

3.2.5.3 Equivalent Noise Charge

Equivalent noise charge (ENC) expresses an amount of noise that appears at the chip input in the absence of useful input signal and is a key chip parameter that affects the energy resolution of the system. Following the standard approach, the total ENC can be divided into three independent components: the white thermal noise associated with the input transistor of the CSA (ENC_{th}), the flicker noise associated with the input transistor of the CSA ($ENC_{1/f}$), and the noise associated with the detector dark leakage current (ENC_{dark}). Noise arising in other components connected to the ASIC input node such as the bias resistor is generally made negligible in a properly designed system. For a first-order shaper, the ENC components can be approximately expressed as:

$$ENC_{th}{}^2 = \left(\frac{8}{3}\right)\frac{kT}{(T_{peck} \times g_m)} \times C_{tot}{}^2$$

$$ENC_{\frac{1}{f}}{}^2 = \left(\frac{K_f}{2}\right) \times WL \times \frac{C_{tot}{}^2}{C_{inp}{}^2}$$

$$ENC_{dark}{}^2 = 2q \times I_{dark} \times T_{peak}$$

$$ENC = \left(ENC_{th}{}^2 + ENC_{\frac{1}{f}}{}^2 + ENC_{dark}{}^2\right)^{\frac{1}{2}}$$

where g_m is the transconductance of the CSA input transistor, C_{tot} is the total capacitance at the input of the CSA, T_{peak} is the shaper peaking time, K_f is the CSA input transistor flicker noise constant, W and L are input transistors width and length, and I_{dark} is the detector leakage current. Note that C_{tot} is the sum of the detector capacitance C_{det}, the gate-source and gate-drain capacitances of the input transistor C_{inp}, and any other feedback or parasitic capacitance at the CSA input originating from the chip package, ESD diodes, and PCB traces. Based on experience with ASIC design and radiation detection module manufacturing, we have assumed C_{inp} to be fractions of pF while the remaining C_{tot} components to be about 1–2 pF. Clearly, particular values are strongly dependent on the chosen technology for the readout IC design and packaging as well as on the chosen connectivity scheme between CZT detector and the chip. It can be easily shown that the optimum peaking time T_{opt} is given by the condition where ENC_{th} is equal to ENC_{dark} leading to the following expression:

$$T_{opt}{}^2 = 4kT\frac{C_{tot}{}^2}{(3g_m q I_{dark})}$$

ENC is typically measured in the lab by measuring the output noise and referring it back to the input using the overall gain of the system. It is also possible to measure channel performance using the scope. By acquiring the channel shaper output signal on the oscilloscope at 1 MHz sampling frequency (e.g.,

5 msec observed time on 5,000 points), a Fast Fourier Transform (FFT) can be applied to the acquired data and the resulting spectrum calculated for each frequency. The noise expressed in mV is calculated as the standard deviation with respect to the shaper output average value and expressed as the equivalent noise in terms of electrons (by considering the known nominal gain of the data channel, typically about in the order of hundreds of mV per fC of the input charge). The result of these calculations is ENC value expressed in number of electrons, typically in a hundreds of electrons range depending what electronics is used, how high the count rate is, and the loading capacitance at the detector input.

3.2.5.4 Signal Shaping

The low-noise amplifier is typically followed by a filter, frequently referred to as the shaper, responding to an event with a pulse of defined shape and finite duration (width) that depends on the time constants and number of poles in the transfer function. The shaper's purpose is twofold: first, it limits the bandwidth to maximize the signal-to-noise ratio; second, it restricts the pulse width in view of processing the next event. Extensive calculations have been made to optimize the shape, which depends on the spectral densities of the noise and system constraints (e.g., available power and count rate).

Optimal shapers are difficult to realize, but they can be approximated, with results within a few percent from the optimal, either with analog- or digital-processors, the latter requiring analog-to-digital conversion of the charge amplifier signal (anti-aliasing filter may be needed). In the analog domain, the shaper can be realized using time-variant solutions that limit the pulse width by a switch-controlled return to baseline or via time-invariant solutions that restrict the pulse width using a suitable configuration of poles. The latter solution is discussed here as it minimizes digital activity in the front-end channels.

In a front-end channel, the time-invariant shaper responds to an event with an analog pulse, the peak amplitude of which is proportional to the event charge, Q. The pulse width, or its time to return to baseline after the peak, depends on the bandwidth (i.e., the time constants) and the configuration of poles. The most popular unipolar time-invariant shapers are realized either using several coincident real poles or with a specific combination of real and complex-conjugate poles. The number of poles, n, defines the order of the shaper. Designers sometimes prefer to adopt bipolar shapers, attained by applying a differentiation to the unipolar shapers (the order of the shaper now is $n - 1$). Bipolar shapers can be advantageous for high-rate applications but at expenses of a worse signal-to-noise ratio.

In typical readout system, the shaping time varies from fraction of μs up to several μs. The shaping time is defined as the time-equivalent of the standard deviation of the Gaussian output pulse. In the laboratory, it is the full width of the pulse at half of its maximum value (FWHM) that is typically being measured. FWHM value is greater than the shaping time by a factor of 2.35.

The DC component of the shaper from which the signal pulse departs is referred to as the output baseline. Since most extractors process the pulses' absolute amplitude, which reflects the superposition of the baseline and the signal, it is important to properly reference and stabilize the output baseline. Non-stabilized baselines may fluctuate for several reasons, like changes in temperature, pixel leakage current, power supply, low-frequency noise, and the instantaneous rate of the events. Non-referenced baselines also can severely limit the dynamic and/or the linearity of the front-end electronics, as in high-gain shapers where the output baseline could settle close to one of the two rails, depending on the offsets in the first stages. In multiple front-end channels sharing the same discrimination levels, the dispersion in the output baselines can limit the efficiency of some channels.

3.2.5.5 Peak Detection

Peak detector is one of the critical blocks in the radiation signal detection system as accurate photon energy is determined by the detected peak amplitude. Standard PDs may be sampled or asynchronous solutions. Sampled PDs are more precise but suffer from high-circuit complexity and high-power dissipation. Asynchronous PDs have simpler structure but suffer from lower output precision.

3.3 Examples of XRD Readout ICs

There are literally hundreds of readout ICs that have been published in the literature. This section summarizes the design and performance characteristics of three photon counting and three spectroscopic devices.

3.3.1 Photon Counting ICs

Fast readout electronic circuits have been developed to reach count rates of several millions counts per second [4–22]. These systems provide coarse energy resolution given by a limited number of discriminators and counters. This section provides some information about the most important photon-counting devices.

3.3.1.1 TIMEPIX and MEDIPIX

MEDIPIX family consists of the several ICs grouped in two families: TIMEPIX and MEDIPIX. TIMEPIX readout IC is the ASIC developed in the framework of the MEDIPIX2 collaboration [25]. The pixel matrix consists of 256×256 pixels with a pitch of 55 μm which gives a sensitive area of about $14 \times 14\,\text{mm}^2$. TIMEPIX is designed in a 0.25 μm CMOS process and has about 500 transistors per pixel. The chip has one threshold and can be operated in photon counting (PC), time over threshold (ToT), or time of arrival (ToA) modes. The principles of the different operating modes are described in detail in the literature [25].

In the photon-counting mode, the counter is incremented once for each pulse that is over the threshold, while for the ToT mode the counter is incremented as long as the pulse is over the threshold. In the time of arrival mode, the pixel starts to count when the signal crosses the threshold and keeps counting until the shutter is closed.

While TIMEPIX is a general purpose chip, the MEDIPIX is aimed specifically at X-ray imaging [25]. It can be configured with up to eight thresholds per pixel and features analog charge summing over dynamically allocated 2×2 pixel clusters. The intrinsic pixel pitch of the ASIC is 55 μm as in TIMEPIX. Silicon die can be bump bonded at this pitch (fine pitch mode) and the chip can be run with either four thresholds per pixel in single pixel mode (SPM) or with two thresholds per pixel in charge summing mode (CSM). Optionally, the chip can be bump bonded with a 110 μm pitch, combining counters and thresholds from four pixels. Operation is possible in SPM with eight thresholds per pixel or in CSM having four thresholds and summing charge of a $220 \times 220\,\mu\text{m}^2$ area.

Being a very versatile and configurable chip, there is also the possibility to utilize two counters per pixel and run in continuous read/write mode where one counter counts while the other one is being read out. This eliminates the readout dead time but comes at a cost of losing one threshold since both counters need to be used for the same threshold. Finally, the charge-summing mode is a very important feature to combat contrast degradation by charge sharing in semiconductors detectors with small pixels.

3.3.1.2 ChromAIX IC

Multi-energy resolving ASIC called ChromAIX has been designed by Philips Corporation to support Spectral CT applications. In order to enable K-edge imaging, at least three spectrally distinct measurements are necessary; for a photon-counting detector the simplest choice is to have at least the same number of different energy windows. With more energy windows, the spectrum of incident X-ray photons is sampled more accurately, thus improving the separation capabilities.

The ChromAIX ASIC accommodates a sufficient number of discriminators to enable K-edge imaging applications. Post-processing allows separating the Photo effect, Compton effect, and one or possibly two contrast agents with their corresponding quantification. The ChromAIX ASIC is a pixelated integrated circuit that has been devised for direct flip-chip connection to a direct converting crystal like CZT. The design target in terms of observed count rate performance is 10 Mcps/pixel, which corresponds to approximately 27.2 MHz/pixel periodic pulses, assuming a paralyzable dead-time model. Although the pixel area in CT is typically about 1 mm², both the ASIC and direct converter features a significantly smaller pixel or sub-pixel. In this way, significantly higher rates can be achieved at an equivalent CT pixel size, while further improving the spectral response of the detector via exploiting the so-called small-pixel effect. The sub-pixel should not be made too small, since charge-sharing effects then start to degrade the spectral performance. Very small pixels would need counter-measures as implemented in Medipix-3, the effectiveness of which at higher rates remains doubtful due to charge-sharing effects.

The ChromAIX ASIC consists of a CSA and a pulse shaper stage, as any other photon-counting device. The CSA integrates the fast transient current pulses generated by the direct converter, providing a voltage step-like function with a long exponential decay time. The shaper stage represents a band-pass filter that transforms the aforementioned step-like function into voltage pulses of a defined height. The height of such pulses is directly proportional to the charge of the incoming X-ray photon. A number of discriminator stages are then used to compare a predefined value (i.e. energy threshold) with the height of the produced pulse. When the amplitude of the pulse exceeds the threshold of any given discriminator, the associated counter will increment its value by one count.

In order to achieve 10 Mcps observed Poisson rates, which would typically correspond to incoming rates exceeding 27 Mcps, a very high band-width is required. The two-stage approach using a CSA and a shaper allows achieving such high rates while relaxing the specification of its components. The design specification in terms of ENC was 400 e-, which corresponds to approximately 4.7 keV FWHM. Simulations of the analogue front-end have been carried out to evaluate the noise performance of the channel. According to these simulations, the complete analogue front-end electronic noise (CSA, shaper, and discriminator input stage) amounts to approximately 2.51 mV$_{RMS}$, which in terms of energy resolution corresponds to approximately 4.0 keV FWHM for a given input equivalent capacitance.

3.3.1.3 PILATUS IC

PILATUS is a hybrid pixel detector system operating in the single-photon counting mode; it was developed at the Paul Scherrer Institut for the needs of macromolecular crystallography at the Swiss Light Source (SLS). A calibrated PILATUS module has been extensively characterized with monochromatic synchrotron radiation. The detector was also tested in surface diffraction experiments at the SLS, whereby its performance regarding fluorescence suppression and saturation tolerance were evaluated, and have shown to greatly improve the sensitivity, reliability, and speed of surface diffraction data acquisition.

The operation of the PILATUS ASIC is as follows. The incident photons are directly transformed into electric charge in the semiconductor sensor, which is transferred via the bump bond to the input of the readout pixel. A schematic of the PILATUS readout chip pixel cell is presented in Figure 3.8. The analog front-end of a readout pixel consists of a charge-sensitive pre-amplifier (CSA) and an AC coupled shaper. The gain and shaping time of the CSA are adjusted with a global voltage (V_{rf}). An analog pulse from the shaper is discriminated against a threshold in the comparator after amplification. The comparator threshold of each pixel is set with a global threshold voltage (V_{cmp}) and is further individually trimmed using an additional in-pixel 6-bit digital-to-analog converter (DAC). If the pulse amplitude exceeds the threshold, a digital signal is produced which increments the 20-bit counter. This detection principle is free of dark current and readout noise effects but requires precise calibration of the pixel threshold for optimum performance.

3.3.2 Spectroscopic ICs

3.3.2.1 IDEF-X

IDeF-X HD is the last generation of low-noise radiation-hard front-end ASICs designed by CEA/Leti for spectroscopy with CZT detectors [23]–[24]. The chip, as shown in Figure 3.9, includes 32 analog channels to convert the impinging charge into an amplified pulse shaped signal and a common part for slow control and readout communication with a controller.

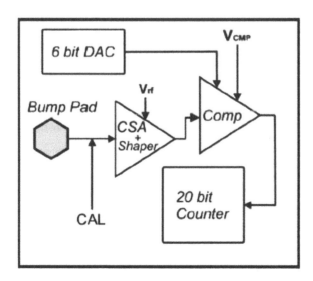

FIGURE 3.8
Architecture of the PILATUS IC readout cell [34].

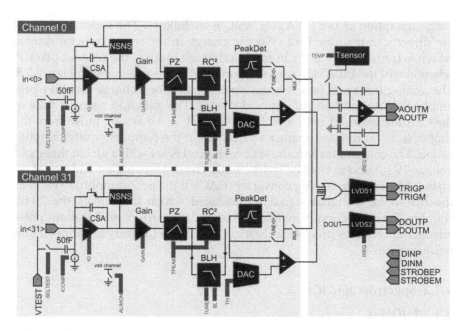

FIGURE 3.9
32-channel IDEF-X IC architecture.

The first stage of the analog channel is a charge-sensitive preamplifier (CSA) based on a folded cascode topology with an inverter input amplifier. It integrates the incoming charge on a feedback capacitor and converts it into voltage; the feedback capacitor is discharged by a continuous reset system realized with a PMOS transistor. The increase of drain current in this transistor during the reset phase is responsible for a non-stationary noise; to reduce the impact of this noise on the equivalent noise charge, a so-called non-stationary noise suppressor was implemented for the first time in this chip version using a low-pass filter between the CSA output and the source of the reset transistor to delay this noise.

The second stage is a variable gain stage to select the input dynamic range from 10 fC (250 keV) to 40 fC (1 MeV). The third stage is a pole zero cancellation (PZ) implemented to avoid long-duration undershoots at the output and to perform a first integration. The next stage of the analog channel is a second-order low-pass filter (RC²) with variable shaping time. To minimize the influence of the leakage current on the signal baseline, a so-called baseline holder (BLH) was implemented by inserting a low-pass filter in the feedback loop between the output of the RC² filter and the input of the PZ stage. The DC level at the output is stabilized for leakage current up to 7 nA per channel. The output of each analog channel feeds a discriminator and a stretcher. The discriminator compares the amplitude with an in-pixel reference low-level threshold to detect events. The stretcher consists of a PD and a storage capacitor to sample and hold the amplitude of the signal which is proportional to the integrated charge and hence to the incident energy. In addition, each channel can be switched off by slow control programming to reduce the total power consumption of the ASIC when using only few channels of the whole chip.

The slow control interface was designed to minimize the number of signals and to get the possibility to connect together up to 8 ASICs and address them individually. This optimization has allowed reducing the electrical interface from 49 pins in Caliste 256 to 16 pins in Caliste-HD for the same number of channels using low-voltage differential signals (LVDS). When an event is detected by at least one channel, a global trigger signal (TRIG) is sent out of the chip. The controller starts a readout communication with 3 digital signals (DIN, STROBE, and DOUT) to get the address of the hit ASIC and then the hit channels. Then the amplitudes stored in the peak detectors of the hit channels are multiplexed and output using a differential output buffer (AOUT). The whole readout sequence lasts between 5 and 20 µs, according to the set delays and clock frequencies and the number of channels to read out.

3.3.2.2 *VAS UM/TAT4*

The VAS UM/TAT4 ASIC chip is used to read-out both the amplitude of charge induction and the electron drift time independently for each anode pixel [27]. This readout IC has 128 channels, each with a charge-sensitive preamp and two CR–RC unipolar shapers with different shaping times. The slow shaper has 1-µs peaking time and is coupled to a peak-hold stage to record pulse amplitude. The fast shaper has a 100-ns shaping time and is coupled to simple level discriminators for timing extraction.

Of the 128 channels, 121 are connected to the pixels, 1 is connected to the grid, and 1 is connected to the cathode. Compared to the anodes, the polarity of the signals is reversed for the cathode and grid. The peak-hold properties, signal shaping, ASIC noise, and triggering procedures are included in the ASIC read-out system model. The fast shaper can trigger off pulses as small as 30 keV for the anode and 50 keV for the cathode. Only the pixels with slow-shaped signals greater than a noise discrimination threshold of 25 keV are used in operation.

VAS UM/TAT4 is particularly well suited for three-dimensional imaging and detection using thick CZT detectors (>10 mm) with high-energy photons (>1 MeV). Three-dimensional position-sensing techniques enable multiple-pixel events of pixelated CZT detectors to be used for 4π Compton imaging. Multiple-pixel events occur by either multiple gamma-ray interactions or charge sharing from a single electron cloud between adjacent pixels. To perform successful Compton imaging, one has to correct for charge sharing. There is a large research effort at University of Michigan under direction of Professor Zhong He to resolve these complicated signal processing issues and to re-construct the trajectory of incoming photons for dirty bomb detection and other high-energy applications that might or might not be relevant to the XRD imaging [28–32].

3.3.2.3 HEXITEC

HEXITEC was a collaborative project between the Universities of Manchester, Durham, Surrey, Birkbeck, and The Science and Technology Facilities Council (STFC) Technology Facilities Council (STFC). The objective of the program was to develop a new range of detectors such as CZT for high-energy X-ray imaging applications. The project has been funded by EPSRC on behalf of RCUK under the Basic Technology Program.

The HEXITEC ASIC consists of a 80×80 pixel array on a pitch of 0.25 mm [26]. Each pixel contains a 52 μm bond pad which can be gold stud bonded to a CZT detector. Figure 3.10 shows a block diagram of the electronics contained in each HEXITEC ASIC pixel. Charge is read from each of the CZT detector pixels using a charge amplifier, which has a selectable range and a feedback circuit which compensates for detector leakage currents up to 50 pA.

The output from each charge amplifier is filtered by a 2 μs peaking circuit comprising a CR–RC shaper followed by a second-order low-pass filter. A peak hold circuit maintains the voltage at peak of the shaped signal until it can be read out. Three track-and-hold buffers are used to sample the shaper and peak hold voltages sequentially prior to the pixel being read.

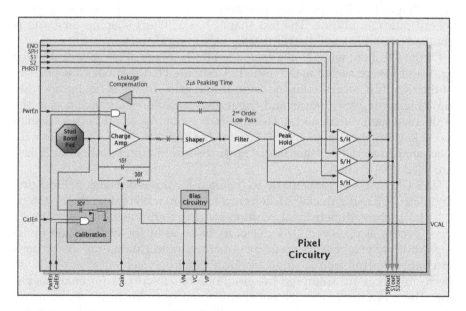

FIGURE 3.10
Block diagram of the HEXITEC architecture.

The HEXITEC is read out using a rolling shutter technique. A row select register is used to select the row which is to be read out. The data from each pixel becomes available on all column outputs at the same time, and at this point, the peak hold circuits in that row can be reset to accept new data. The data being held on the column output is read out through a column multiplexer. The column readout rate is up to 25 MHz and the total frame rate depends on the number of pixels being read out. The main limitation of the HEXITEC is a maximum count rate due to 10 kHz frame readout scheme.

3.4 Readout IC Operational Issues

3.4.1 Threshold Equalization

As part of the pixel detector calibration, the ASIC chip has to be equalized in order to minimize the threshold dispersion between pixels. This requirement results from the fact that the threshold that the pixel sees is applied globally but the offset level of the pixel can be slightly different due to process variations affecting the baseline of the preamplifier.

The equalization is performed with a threshold adjustment DAC in each pixel. The resolution of the adjustment DAC is usually in the range of 4 bits depending on particular ASIC implementation. The standard way to calculate the adjustment setting for each pixel this is by scanning the threshold and finding the edge of the noise, then aligning the noise edges. This adjusts correctly for the offset level of the pixel but gain variations can still deteriorate the energy resolution at a given energy. To correct for the gain mismatch, either test pulses or monochromatic X-ray radiation has to be used for the equalization. Equalizing at the energy of interest instead of the zero level might be also preferred.

3.4.2 Energy Calibration

Depending on the ASIC architecture, there are two types of energy calibration that needs to be done: calibration of the threshold and calibration of the time over threshold response (if applicable). For photon-counting chips as MEDIPIX, the only calibration required is the threshold response while in time over threshold readout ICs such as TIMEPIX the ToT response has to be calibrated as well. Virtually all spectroscopic ICs need to undergo energy calibration procedures as electronics circuitry basically has no understanding of photon energy expressed in keV.

To calibrate the threshold, we need monochromatic photons or at least radiation with a pronounced peak. These can be obtained from radioactive sources, by X-ray fluorescence, or from synchrotron radiation like Am241 and/or Co57 point sources. To find the corresponding energy for a certain threshold, the threshold is scanned over the range of the peak obtaining an integrated spectrum. The data is then either directly fitted with an error or sigmoid function or first differentiated and then fitted with a Gaussian function. From this fit, the peak position and energy resolution can be extracted. Repeating the procedure for multiple peaks the result can then be fitted with a linear function, and the relationship between voltage threshold setting and deposited energy in the detector is found.

3.4.3 Charge Sharing Corrections

Charge sharing between pixel is a highly detrimental effect in XRD signal detection as explained in Chapter 2. To illustrate the techniques used for charge sharing correction, it is helpful to review charge sharing using the diagrams shown in Figure 3.11.

Suppose that a pixelated detector is divided into 9 pixels as shown in Figure 3.11a. If a X-ray photon strikes pixel 5 near the center as in Figure 3.11b, then most of the charge will be localized to that pixel and very little charge sharing will occur. However, if the photon lands near the pixel boundaries as in Figure 3.11c, then the charge will be shared amongst pixels 1, 2, 4, and 5; low energy events will be detected in the pixels 1, 2, and 4, creating a low energy tail in the spectrum. To correctly detect the energies of the incoming photons, the system must recognize that a single photon event has spread its charge in a cluster of pixels, determine where the photon has most likely landed (usually the pixel with the largest charge deposition), and assign all surrounding charge to that single pixel.

In general, when the pixel size is starting to approach the size of the charge cloud, the input signal is subjected to charge sharing. Charge sharing creates a characteristic low energy tail and leads to a reduced contrast and distorted spectral information. To counteract this problem, there are two possibilities, either to use larger pixels (reduced spatial resolution) or to implement charge summing on a photon by photon basis.

For lower rates and with detectors that store the energy information in each pixel either using ToT (TIMEPIX) or a peak-and-hold circuit (HEXITEC), the

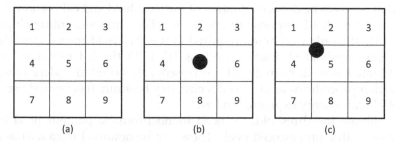

FIGURE 3.11
Charge sharing in a pixelated detector.

charge summing can be done offline. However, this requires that you do not have a second hit in the same pixel before you read the first one out. Using this approach, you also lose charge that is below the detection threshold.

With a photon near the pixel boundaries as in Figure 3.12a, most of the charge is collected by the center pixel and some of the charge is shared with adjacent pixels. Some readout ICs are able to recognize that these events are coincident in time and reassign the charge to the correct pixel. The MEDIPIX readout IC from CERN was the first ASIC to implement this algorithm called the *Charge Summing Mode* (CSM), where the analog charge can be summed up in a 2×2 cluster before being compared to the threshold. MEDIPIX IC also has a mode in which this functionality is disabled, called the *Single Pixel Mode* (SPM).

The advantage of this approach is that it can handle much higher interaction rates and that even charge below the threshold is summed as long as one pixel is triggered. However, since this correction has to be implemented in the ASIC architecture, it complicates the chip design and is less flexible when interfacing with various interconnects.

3.4.4 Pile-Up Effects

Given that the processing of each photon takes time, there will be problems with pileup effects at high count rates. Pileup happens when a second photon arrives in the same pixel before the first one is processed. Depending on the system architecture, the second photon could either be lost or added to the signal of the first photon. The result will be a deviation from linear behavior for the count rate. This deviation can be corrected for up to a certain limit in photon-counting devices, but more problematic are the spectral distortions due to pileup that cannot be corrected for. For this reason, the operation of spectroscopic ASIC is limited to the maximum count rate that is not causing any pileup effects. Different detectors will have different responses and it is important that the detector is characterized and is suitable for the flux in a specific application. Since the flux is measured per area, smaller pixels offer an advantage of a smaller number of photons per second per pixel.

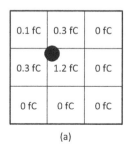

(a)

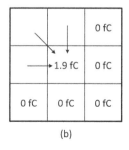

(b)

FIGURE 3.12
Charge sharing correction in a pixelated detector.

The X-ray photon events in the CZT detector occur randomly following a Poisson distribution. The probability density function between successive events is given by [5]:

$$p(t) = r \cdot \exp(-rt)$$

where r is the average incoming photon rate. While pile-up occurs, there will be a deviation from the linear relationship between flux rate and count rate; as illustrated by the PILATUS curves shown in Figure 3.13, the count rate in the ASIC begins to saturate as the flux rate increases beyond a certain limit.

An equivalent way of characterizing this limitation is via a parameter called dead time, which is the minimum amount of time that must separate photon arrivals for them to be counted separately. If successive photons arrive within this dead time window, paralysis occurs in the readout IC count as shown in Figure 3.14.

As illustrated in Figure 3.14, a single photon generates a pulse which is counted correctly. Subsequently, two photons arrive within the dead time window and is only counted as one event. Finally, multiple photons arrival causes pulse overlaps such that the pulse shaper output does not fall below the threshold and paralysis occurs in the IC count.

Modern ICs combat this problem by introducing non-paralyzable counting modes. An example of the implementation is the PILATUS3 Instant Retrigger Architecture. This is accomplished by re-evaluating the pulse shaper output after a programmable amount of time. If the output is still above threshold, it is assumed that pileup has occurred and the counting circuit is retriggered. This technique prevents the count rate from freezing due to multiple photon arrivals.

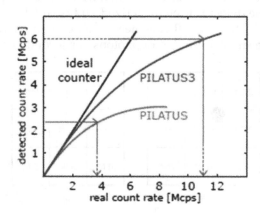

FIGURE 3.13
PILATUS3 ASIC count rate vs. incoming flux rate [33].

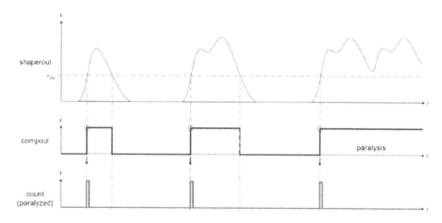

FIGURE 3.14
Signal waveforms illustrating paralyzable counting [33].

3.4.5 Dark Current Correction

Even when no photons strike the detector, each pixel can still produce some charge due to leakage currents from the detector and the ASIC itself. Hence a calibration procedure is performed at start-up called *dark current correction*. With no X-ray source and the detector in the "dark," the value of each pixel is read and stored into memory. When the system is running, this dark current value is subtracted from the raw value of each pixel, hence removing the artifacts due to the leakage currents. It is also possible to correct for dark current contribution by re-designing analog front-end so the correction happens simultaneously with the photon detection.

An example of one of the dark current schemes is shown in Figure 3.15. After dark current correction, each pixel is compared to a programmable threshold to determine whether a valid event has occurred. This process is

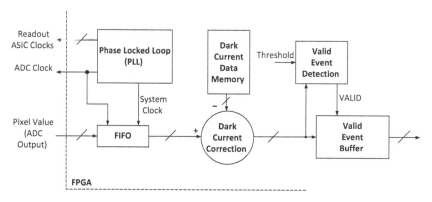

FIGURE 3.15
Dark current correction and valid event detection.

used to further remove any noise from the data. A pixel value that is above the threshold is called a *valid event* for the discussion below.

3.5 Conclusions

Currently, most digital radiation detectors for security applications are based on integrating the X-ray quanta (photons) emitted from the X-ray tube for each frame. This technique is vulnerable to noise due to variations in the magnitude of the electric charge generated per X-ray photon. Higher energy photons deposit more charge in the detector than lower energy photons so that in a quantum integrating detector, the higher energy photons receive greater weight. This effect is undesirable in many detection applications because the higher part of the energy spectrum provides lower differential attenuation between materials, and hence, these energies yield images of low contrast.

Direct conversion X-ray quantum counting detectors solve the noise problem associated with photon weighting by providing better weighting of information from X-ray quanta with different energies. In an X-ray quantum counting system, all photons detected with energies above a certain predetermined threshold are assigned the same weight. Adding the energy windowing capability to the system (i.e., counting photons within a specified energy range) theoretically eliminates the noise associated with photon weighting and decreases the required X-ray dosage by up to 40% compared to integrating systems.

Direct conversion detectors are also essential to XRD detection and imaging applications where precise information about the energy of incoming photons is crucial. This chapter reviews key challenges that are present in spectroscopic and photon-counting electronics for XRD. Energy resolution, charge-sharing correction, high flux capability, and pileup effects affect both the count rate linearity and the spectral response. One way to counteract that problem is to use smaller pixels, but smaller pixel will lead to more charge sharing. In this respect, the various chips offer an interesting combination of relatively small pixels and still very good energy resolution.

The readout integrated circuits (ROIC) design issues have been discussed in detail in this chapter. A summary classification of those ICs is shown in Figure 3.16. The ROIC family can be divided into spectroscopy and photon-counting chips. The photon-counting chips usually operate in a synchronous manner while the spectroscopic chips can be both asynchronous and synchronous. Although over 100 ROIC have been designed and used, it is no doubt that new designs will enter the marketplace in the next 5–10 years and are likely to be based on very advanced CMOS chip fabrication technologies.

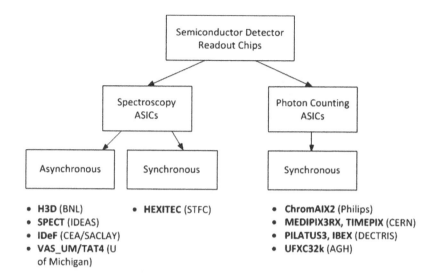

FIGURE 3.16
Classification of the readout ICs discussed in this chapter.

References

1. *Source*: https://en.wikipedia.org/wiki/X-ray.
2. *Source*: http://amptek.com/products/gamma-rad5-gamma-ray-detection-system/#7.
3. *Source*: http://www.i3system.com/eng/n_tech/tech2.html.
4. R. Redus, *Charge Trapping in XR-100-CdTe and -CZT Detectors*, Amptek, 2007.
5. R. Macdonald, Design and implementation of a dual-energy X-ray imaging system for organic material detection in an airport security application, *Proc. of SPIE Vol. 4301*, 2001.
6. Y. A. Boucher, F. Zhang, W. Kaye, and Z. He, *Study of Long-Term CdZnTe Stability using the Polaris System*, IEEE NSS-MICS, 2012.
7. H. Chen, F. Harris, S. Awadalla, P. Lu, G. Bindley, H. Lenos, and B. Cardoso, *Reliability of Pixellated CZT Detector Modules Used for Medical Imaging and Homeland Security*, SPIE Invited Paper 2012.
8. G. Harding, H. Strecker, S. Olesinski, and K. Frutschy, Radiation source considerations relevant to next-generation X-ray diffraction imaging for security screening applications, edited by F. Patrick Doty, H. Bradford Barber, Hans Roehrig, Richard C. Schirato, *Proc. of SPIE Vol. 7450, Penetrating Radiation Systems and Applications*, 2009.
9. S. Skatter, H. Strecker, H. Fleckenstein, and G. Zienert, *CT-XRD for Improved Baggage Screening Capabilities*, IEEE NSS MICS Conference, 2009.
10. G. Harding, H. Strecker, and P. Edic, Morpho detection and GE global research, high-throughput, material-specific, next-generation X-ray diffraction imaging (XDi), *IEEE NSS-MICS Conference 2010*.

11. V. Rebuffel, J. Rinkel, J. Tabary, and L. Verger, New Perspectives of X-ray Techniques for Explosive Detection Based on CdTe/CdZnTe Spectrometric Detectors, *International Symposium on Digital Industrial Radiology and Computed Tomography*, Berlin, 2012.

12. Y. Tomita, Y. Shirayanagi, S. Matsui, M. Misawa, H. Takahashi, T. Aoki, and Y. Hatanaka, *X-ray Color Scanner with Multiple Energy Differentiate Capability*, 0-7803-8701-5/04, 2004 IEEE.

13. A. A. Faust, R. E. Rothschild, P. Leblanc, and J. Elton McFee, Development of a coded aperture X-ray backscatter imager for explosive device detection, *IEEE Transactions on Nuclear Science* 56(1) (Feb. 2009) 299–307.

14. G. Harding, and B. Schreiber, Coherent X-ray scatter imaging and its applications in biomedical science and industry, Philips Research, *Radiation Physics and Chemistry* 56 (1999) 229–245.

15. Online. www.morpho.com/IMG/pdf/Xray_Diffraction_Technology_Presentation.pdf.

16. S. D. M. Jacques, C. K. Egan, M. D. Wilson, M. C. Veale, P. Seller, and R. J. Cernik, *A laboratory system for element specific hyperspectral X-ray imaging.*

17. E. Cook, R. Fong, J. Horrocks, D. Wilkinson, and R. Speller, Energy dispersive X-ray diffraction as a means to identify illicit materials: A preliminary optimisation study, *Applied Radiation and Isotopes* 65 (2007), 959.

18. L. Verger, E. Gros d'Aillon, O. Monnet, G. Montemont, and B. Pelliciari, New trends in-ray imaging with CdZnTe/CdTe at CEA-Leti, *Nuclear Instruments and Methods A*, 571 (Feb. 2007), 33–43.

19. J. Iwanczyk, E. Nygard, O. Meirav, J. Arenson, W. Barber, N. Hartsiugh, N. Malakhov, and J. C. Wessel, Photon counting energy dispersive detector arrays for X-ray imaging, in *Proc. IEEE Nucl. Sci. Symp. Rec.*, 2007, pp. 2741–2748.

20. C. Szeles, S. Soldner, S. Vydrin, J. Graves, and D. Bale, CdZnTe semiconductor detectors for spectrometric X-ray imaging, *IEEE Transactions on Nuclear Science*, 55 (Feb. 2008), 572–582.

21. S. Mikkelsen, D. Meier, G. Maehlum, P. Oya, B. Sundal, and J. Talebi, An ASIC for multi-energy X-ray counting, in *Proc. IEEE Nucl. Sci. Symp. Rec.*, 2008, pp. 294–299.

22. O. Tümer, V. Cajipe, M. Clajus, S. Hayakawa, and A. Volkovskii, Multi-channel front-end readout IC for position sensitive solid-state detectors, in *Proc. IEEE Nucl. Sci. Symp. Rec.*, 2006, pp. 384–388.

23. X. Wang, D. Meier, B. Sundal, B. Oya, P. Maehlum, G. Wagenaar, D. Bradley, E. Patt, B. Tsui, and E. Frey, A digital line-camera for energy resolved X-ray photon counting, in *Proc. IEEE Nucl. Sci. Symp. Rec.*, 2009, pp. 3453–3457.

24. Brambilla, C. Boudou, P. Ouvrier-Buffet, F. Mougel, G. Gonon, J. Rinkel, and L. Verger, Spectrometric performances of CdTe and CdZnTe semiconductor detector arrays at high X-ray flux, in *Proc. IEEE Nucl. Sci. Symp. Rec.*, 2009, pp. 1753–1757.

25. J. Rinkel, G. Beldjoudi, V. Rebuffel, C. Boudou, P. Ouvrier-Buffet, G. Gonon, L. Verger, and A. Brambilla, Experimental evaluation of material identification methods with CdTe X-ray spectrometric detector, *IEEE Transactions on Nuclear Science*, 58(2) (Oct. 2011), 2371–2377.

26. L. Tlustos, Spectroscopic X-ray imaging with photon counting pixel detectors, *Nuclear Instruments and Methods A*, 623(2) (Nov. 2010), 823–828.

27. L. Jones, P. Seller, M. Wilson, A. Hardie, HEXITEC ASIC—a pixellated readout chip for CZT detectors, *Nuclear Instruments and Methods A*, 604 (2009), 34–37.

28. F. Zhang and Z. He, New readout electronics for 3-D position sensitive CdZnTe/HgI2 detector arrays, *IEEE Transactions on Nuclear Science*, 53(5) (2006), 3021–3027.

29. Online. Radiography (Plain x-rays): www.medicalradiation.com/types-of-medical-imaging/imaging-using-x-rays/radiography-plain-x-rays/.

30. J. Scampini, *Introduction to Computed Tomography (CT)*, Maxim Integrated, 2010. [Online]. Available: www.maximintegrated.com/en/app-notes/index.mvp/id/4682.

31. Morpho Detection, *XDi - the Ultimate Automatic Type D Liquid Explosives Detection System for Checkpoints*, Hamburg, Germany, 2014.

32. R. Ballabriga, *Chips Developed at CERN in the Framework of the Medipix3 Collaborations*, Geneva, 2013.

33. R. Ballabriga, *Review of Hybrid Pixel Detector Readout ASICs for Spectroscopic X-Ray Imaging*, 2016.

34. C. Schulze-Briese, *The New PILATUS3 ASIC with Instant Retrigger Technology*, in PIXEL 2012, Inawashiro, Japan, 2012.

4

Applications of X-Ray Diffraction Imaging in Medicine

Manu N. Lakshmanan

National Institutes of Health, Clinical Center

CONTENTS

4.1 Introduction and Background

After work by Friedrich et al. (1913) and Bragg & Bragg (1913) demonstrating the principles and utility of XRD, it has had tremendous impact in the field of biology through reconstructing the structures of biological molecules and macromolecules primarily through X-ray crystallography, leading to Nobel Prizes in Chemistry and Medicine in over 10 different years (e.g., see Perutz et al. 1965; Watson et al. 1953; Blundell et al. 1972; Deisenhofer et al. 1985; Rosenbaum et al. 2009). However, due to the necessity in X-ray crystallography that the sample have a highly organized crystal structure, the utility of XRD initially could not be extended to medical imaging, where the samples, such as human soft tissue, are amorphous. Work by Debye & Scherrer (1916) and Hull (1917) demonstrated the concept of powder diffraction, in which XRD patterns can be measured for the identification of non-crystallized samples as well, opening the door for XRD applications in medical imaging. Subsequently, powder diffraction patterns have been widely measured and analyzed for medical imaging applications.

XRD imaging (i.e., spatially resolved reconstructions of XRD signal) was first proposed for applications in medical imaging by Harding et al. (1987) and Kosanetzky et al. (1987), who pointed out that the diffraction patterns from body fat, muscle, and bone were sufficiently distinct that they could be differentiated solely based on XRD measurements (see Figure 4.1(a)), and that these diffraction patterns could be spatially resolved in the sample using coherent scatter computed tomography (CSCT) architecture (Figure 4.1(b) and (c)). Figure 4.1 shows the XRD imaging concept and the contrast

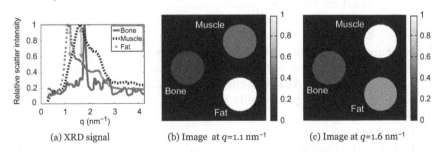

(a) XRD signal   (b) Image at q=1.1 nm^{-1} (c) Image at q=1.6 nm^{-1}

FIGURE 4.1

(a) The XRD patterns for three biological materials measured by Harding et al. (1987). The x-axis shows the momentum transfer, q, an X-ray parameter related to the X-ray energy and scattering angle, $q \equiv Es$ in $(\theta/2)/hc$. The contrast in scatter signal between the three types of tissue at certain q values (e.g., the contrast highlighted with green arrows between muscle and fat at the q locations of their peaks) can enable differentiation and material identification of those tissues in the reconstructed XRD images. If the three tissues were placed ex vivo in test tubes, then (b) and (c) show a schematic of XRD images at the q values indicated by the green arrows in (a) where fat and muscle can be best differentiated. As expected based on (a), in the XRD images in (b) and (c) fat is most intense at $q = 1.1$ nm^{-1} and muscle is most intense at $q = 1.6$ nm^{-1}.

that can be achieved between muscle and fat—two types of soft-tissue between which there is limited contrast available in transmission-based X-ray imaging—at the two momentum transfer (q) values where the scatter intensities for the two tissues differ the most. In the nearly three decades, since the introduction of the concept of XRD imaging for medical applications, it has been demonstrated for numerous other clinical and diagnostic applications, which will be reviewed in this chapter.

4.2 Osteoporosis Detection: Bone Imaging

4.2.1 Diagnostic Imaging Techniques Used in Current Clinical Practice

Osteoporosis is the most common bone disorder in North America (Grampp et al. 1997). It is characterized by decreased bone mass through the loss of minerals in the bone. Since the first noticeable symptoms for osteoporosis occur only after the bone mass has deteriorated to the point where the skeletal system becomes highly vulnerable to mechanical stresses, it can often lead to serious injury in patients, such as the collapse or fracture of bones, prior to diagnosis and treatment (Grampp et al. 1997). Therefore, the detection of the onset of osteoporosis can reduce risks of serious injury.

The standard technique for assessing osteoporosis is dual energy X-ray absorptiometry (DEXA) (Royle et al. 1999), which is able to detect the decay in bone mineral content that is characteristic of osteoporosis in the cortical bone. However, DEXA is less effective at detecting bone demineralization in the other major type of bone: cancellous bone (Hall et al. 1987). Cancellous bone is less dense than cortical bone, but its structural purpose in the body is significant as it makes up the vertebral bodies in the spine. Moreover, osteoporosis occurs sooner in the cancellous bone in the vertebral bodies than in the cortical bone because cancellous bone is more metabolically active (Hall et al. 1987). Therefore, the detection of osteoporosis in the cancellous bone comprising the vertebral bodies in the spine is more critical than in the cortical bone.

Batchelar et al. (2006) point out that DEXA and quantitative computed tomography (QCT)—another diagnostic imaging technique that has been studied for osteoporosis diagnosis—assess the risk and progression of osteoporosis by providing measurements of the bone mineral density (BMD). They argue so because the BMD is a bulk measure of material content, it does not capture the component of fracture risk from deterioration of the bone structure (e.g., changes in the gross bone size and trabecular microarchitecture) in osteoporosis. Therefore, the BMD does not fully describe bone strength nor the risk of fracture.

Westmore et al. (1997) and Laii & Cunningham (1998) have shown that radiographic absorptiometry and QCT provide greater error due to the variable fat content in the bone. This susceptibility to error becomes more significant for osteoporosis risk assessment because the relative proportion of fat and lipid levels in the bones increase with the onset of osteoporosis, which has been linked to deterioration in bone microarchitecture (Wasnich 1996). Therefore, there is a need for a more direct measure of bone mineral content than the bulk BMD in order to more accurately assess the progression of osteoporosis in the bone.

4.2.2 XRD Imaging for Osteoporosis Diagnosis

The bone XRD pattern was confirmed to be distinct in the previously described study by Harding et al. (1987) and is shown in Figure 4.1(a). Normal healthy bone has sharp XRD rings due to the presence of hydroxyapatite crystals in the bones (Carlström & Finean 1954). In contrast, osteoporosis causes bones to become more diffuse (Carlström & Finean 1954), due to hydrated collagenous tissue making up a greater percentage of the bone after demineralization. This change in structure accompanying osteoporosis that leads to the bone structure becoming more diffuse will result in a corresponding change in the XRD pattern, causing it to appear more like the diffraction patterns from soft tissue, e.g., more like the XRD pattern for fat or muscle in Figure 4.1(a). Therefore, bone demineralization disease can be assessed using XRD by exploiting the differences in XRD patterns between the mineral and fat components of bone. Consequently, XRD measurements can potentially be used for diagnosis of osteoporosis before the patient suffers a broken bone or fracture. As an assessment of the ability of XRD imaging to assess bone content with similar efficacy as QCT, Royle et al. (1999) and Newton et al. (1992) showed that bone mineral density estimates from XRD imaging were highly correlated ($r = 0.95$) with those of QCT.

Royle & Speller (1991) applied powder XRD analysis by obtaining the energy-dispersive diffraction spectra in the bone from specific regions, such as the bone marrow and bone tissue, and were able to obtain a more sensitive and accurate analysis of the changes in the bone that accompany osteoporosis than what was possible using either DEXA or QCT, which both rely on the bulk measurement of the BMD. In another study, the authors later showed (Royle & Speller 1995) that taking the ratio of the bone tissue XRD peak intensity to that of the bone marrow within the bone can serve as a measure of bone content that correlates with the osteoporotic state of the bone ($r = 0.92$).

Another diagnostically relevant metric from powder XRD signal was introduced by Speller (1999). They showed that by combining measurements of Compton scattering and XRD imaging from bone, the elastic modulus can be computed, which is a measure of bone strength under stress. They were able to show that this elastic modulus computed from XRD imaging and Compton scatter measurements had a higher correlation with the bone's resistance to fracture or provided an improvement in fracture prediction than the metrics (e.g., the BMD) that could be computed using DEXA alone.

Batchelar et al. (2006) began applying XRD imaging using the CSCT architecture for bone disorder diagnosis based on the importance of measuring the spatial structure of bone and its composition. They were able to show that XRD imaging can provide volumetric density estimates of various bone components that can be used to understand bone metabolism beyond the bulk BMD. Specifically, XRD imaging is able to investigate spatial density estimates of hydroxyapatite and collagen in grams per cm^3 with a better than 5% accuracy and better than 4% precision. The XRD density images could be used for the separation of the cancellous and cortical bones (Batchelar & Cunningham 2002). Through focusing on the changes in density of specifically the cancellous bone, which DEXA is unable to accurately assess as discussed earlier, XRD imaging can potentially detect the onset of osteoporosis in them or in metabolically active vertebral bodies of the spine, where osteoporosis has been shown to occur earlier than the cortical bone.

In another study, Batchelar et al. (2000) showed that XRD imaging can be used to further decompose the volumetric density distributions in the bone down to its collagen and hydroxyapatite compositions. This ability to estimate the collagen and hydroxyapatite compositions can be used as a substitute for bone biopsy in the detection of osteomalacia (Bonucci 1998), a related disease to osteoporosis. Osteomalacia comes about due to deficiency of vitamin D, which causes the bone to be poorly mineralized as in osteoporosis. Its diagnosis requires biopsy because it cannot be detected using DEXA or QCT (Bonucci 1998). Consequently, the application of XRD imaging can potentially spare osteomalacia patients of the invasiveness that would accompany a biopsy.

4.3 Breast Cancer Imaging

Women in the United States have a greater than 12% chance of developing breast cancer during their lifetimes; and just under 14% of breast cancer cases result in death (Desantis et al. 2016). Among cancers, breast cancer is the second leading cause of death in women in the United States (Desantis et al. 2016). Thus, the diagnosis and treatment of breast cancer is one of the most important areas of modern medical research.

4.3.1 Diagnostic Imaging Techniques Used in Current Clinical Practice

The first line of defense (i.e., screening) of breast cancer detection in clinical practice is currently performed using screening mammography, in which the woman's breast is compressed between two paddles and an X-ray radiograph (i.e., the X-ray source is not rotated as in CT) is taken of the compressed breast. Since 2000, greater than 66% of women (even symptom-free healthy women) older than 40 reported having a mammogram during the last two years (Desantis et al. 2016). Despite it by far being the most frequently applied diagnostic imaging tool for breast cancer detection, there are many severe limitations of mammography.

The physics of X-ray transmission imaging of any form (e.g., CT, radiography, or mammography) is largely unsuited to the task of breast cancer detection. Measurements of the X-ray attenuation coefficient for different breast tissues at 18 keV—the characteristic X-ray peak in the molybdenum source spectrum used in mammography—show that while the differences between adipose (fatty) tissue and cancerous breast tissues is ~49% (Johns & Yaffe 1987), the corresponding difference between fibrous (the major component of the breast in women with dense breasts) and cancerous breast tissues is ~5% (Johns & Yaffe 1987). Similarly, calculations of the theoretical limit on the contrast that can be generated from the linear attenuation coefficients between adipose and cancerous tissue is 23% and between fibroglandular and cancerous tissue is 3% (Kidane et al. 1999). Given that X-ray transmission imaging cannot detect differential attenuation differences less than 5% in most clinical imaging situations (Speller 1999), X-ray transmission imaging in most cases therefore cannot visualize breast cancer lesions in patients whose breasts contain considerable amounts of fibroglandular tissues, which is the case for women with dense breasts.

This physics-based analysis of the shortcomings of transmission-based X-ray imaging is further supported by observations of clinical outcomes. Experience in hospitals has shown that mammography is non-specific as nearly 19% of the women who are recalled for further assessment due to having a suspicious mammography are ultimately shown to be free of cancer (Jacobsen et al. 2015). Many of these women who are recalled following a suspicious mammogram must undergo painful invasive biopsies to confirm the absence of cancer. In addition to mammography being non-specific, it also misses over 8% of breast cancers, as its sensitivity has been shown to be just below 92% (Jacobsen et al. 2015).

4.3.2 XRD Patterns for Healthy and Diseased Breast Tissue

The first demonstration of X-ray diffraction for soft-tissue contrast was carried out by Evans et al. (1991) who measured the scatter intensity as a function of angle from exposure to a collimated quasi-monochromatic beam in blood and several types of freshly excised room-temperature breast tissue samples: adipose, fibroglandular, carcinoma, fibrocystic disease and fibroadenoma. They saw significant differences in peak positions between adipose and fibroglandular tissue, with the two resembling oil and water, respectively, in their peak scatter angle locations. They also observed that adipose tissue could be separated from the other soft-tissues based on its peak width and normalized peak intensity.

Explanations of the considerably different XRD properties of cancerous breast tissue have been speculated to be due to the role of collagen, which exhibits strong diffraction patterns (Bear 1942). Collagens make up the extra-cellular matrix in human tissue, which is degraded and penetrated by the presence of cancer (Schönermark et al. 1997). The collagen structure has been shown to be substantially disrupted by malignancy (Raymond & Leong 1991; Pucci-Minafra et al. 1998; Pucci-Minafra et al. 1993). Consequently, the relationship between collagen and cancer, and the strong XRD properties of collagen lead to differences arising between cancerous and healthy breast tissue.

Kidane et al. (1999) performed an exhaustive study and detailed analysis of breast XRD patterns, measuring them from the excised breast tissue samples of 100 patients and comparing the results to histological analysis (see Figure 4.2). They found a factor of two difference in peak form factor intensity between adipose and non-adipose tissues. For the case of differentiating cancerous and fibroglandular tissue, they observed greater XRD intensity at lower momentum transfer values for fibroglandular tissue relative to cancerous. This difference at lower momentum transfer is likely due to the reduced fat content in cancerous tissue, which results in it having a lower intensity at the location of the peak in the adipose tissue form factor, which is at 1.12 nm^{-1}.

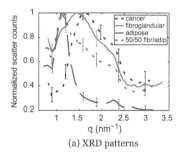
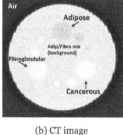
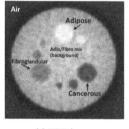

(a) XRD patterns   (b) CT image   (c) XRD image

FIGURE 4.2
(a) The XRD signal for the four types of tissue in the breast, as measured by Kidane et al. (1999) using 100 tissue samples. The considerable difference (19% contrast at 1.1 nm^{-1}) between cancerous and fibroglandular tissue enables XRD imaging to provide more consistent and accurate classification of healthy and cancerous tissue than possible using conventional CT, which provides 3% to 5% contrast between cancerous and fibroglandular tissue. Subfigures (b) and (c) show a comparison of the contrast between breast tissues that can be obtained using CT and XRD imaging (CSCT architecture) using an 8 cm phantom containing different types of breast tissue. The circles of cancerous tissue (bottom right) cannot be differentiated from the circles of fibroglandular tissue (left) using CT; however, the fibroglandular and cancerous circles have more substantially contrasting intensity values in XRD imaging. Subfigures (b) and (c) reproduced from Ghammraoui et al. (2015) by permission of SPIE. All rights reserved. ©SPIE.

A number of subsequent studies have confirmed the ability to use XRD patterns to differentiate cancerous breast tissue from healthy tissue: Lewis et al. (2001), Poletti et al. (2002), Castro et al. (2005), Changizi et al. (2005), Changizi et al. (2008) and Sidhu et al. (2011). Furthermore, a number of studies have shown for at least 30 samples the ability to use discriminant analysis applied to XRD data to accurately classify healthy and malignant breast tissue. These studies have shown sensitivities ranging from 54% to 100%, specificities ranging from 82.3% to 100%, and accuracies ranging from 75% to 97%, which are detailed in Table 4.1. (The relatively poor sensitivity of 54% found by Ryan & Farquharson (2007) is likely due to the aspect of their study that separates theirs from the other studies: their strategy to incorporate Compton scatter measurements along with XRD measurements in making the classification decisions). Additionally, Oliveira et al. (2008) studied the possibility of differentiating benign and malignant tumors as well and achieved 78.6% sensitivity and 62.5% specificity using XRD patterns.

The strength of XRD measurements for breast-cancer classification that is demonstrated from the aforementioned studies and Table 4.1 can also be compared to the physics-based analysis of X-ray transmission-based imaging that was presented at the beginning of the previous subsection (Section 3.1). Whereas previously mentioned, the potential contrast based on X-ray attenuation between fibroglandular and cancerous tissue has been shown to be ~3% to 5% (Johns & Yaffe 1987; Kidane et al. 1999) when measured at the X-ray energies used in mammography, the corresponding potential contrast based on XRD patterns between fibroglandular and cancerous tissue is 19% if measured for example at a momentum transfer value of 1.1 nm^{-1}(see Figure 4.2). Applying this classification ability and potential contrast from the XRD signal between healthy and cancerous tissue, a number of groups have developed XRD imaging for breast cancer imaging, which will be reviewed next.

TABLE 4.1

Quantitative Evaluation of Discriminant Analysis Applied to Classifying Healthy and Cancerous Breast Tissues Based on XRD Patterns

Study	Sensitivity (%)	Specificity (%)	Accuracy (%)
Cunha et al. (2006)	100	95	97
Ryan & Farquharson (2007)	54	100	N/A
Oliveira et al. (2008)	95.6	82.3	75
Conceiçãco et al. (2009)	83	100	N/A
Elshemey et al. (2010)	78 to 94	94 to 100	86 to 97
Elshemey et al. (2013)	82 to 100	91 to 95	85 to 85

The difference between the last two studies, which were carried out by the same authors, is that the breast samples were dehydrated in the latter in order to make the samples more closely resemble breast tissues that are analyzed ex vivo using pathological techniques, and in order to reduce the dominant effect of scattering from water

4.3.3 XRD Imaging of Breast Cancer

Ghammraoui & Badal (2014) show that XRD imaging using the CSCT architecture can characterize tissue composition inside a whole breast. They further demonstrated that XRD imaging provides greater contrast between cancerous and fibroglandular tissue than conventional X-ray CT (Ghammraoui et al. 2015) (see Figure 4.2(b) and (c)). In that study, they also used the contrast-to-noise ratio metric, or CNR, to demonstrate that the pencil beam CSCT design outperforms their cone-beam CSCT design. Ghammraoui et al. (2016) incorporated a maximum likelihood estimate of scatter components algorithm that reconstructed XRD images from material basis functions (e.g., using basis materials such as cancerous, adipose tissues, etc.) to demonstrate that XRD imaging can separate adipose tissue and water using radiation doses comparable to those used in conventional breast CT. They were also able to show that the CSCT XRD imaging architecture has sufficiently fine resolution to potentially image microcalcifications in the breast; which are one of the earliest presenting features of breast cancer (Gülsün et al. 2003), are present in 80% of breast cancers, and are generally less than 0.5 mm in size and can be microscopic (Chan et al. 1987).

4.3.4 XRD Imaging of Lumpectomy Specimens: CSCT Approach

Other than imaging whole breasts, another potential clinical application of XRD imaging that exploits its ability to detect breast cancer is intra-operative margin detection. One of the options available to breast cancer patients is to receive a breast conserving surgery (BCS) (a.k.a. lumpectomy) in which only the breast tumor is removed rather than the entire breast. In order to ensure the success of a lumpectomy procedure, the surgeon must confirm that the entire breast tumor was removed from the patient by identifying that there is a margin of healthy tissue around the tumor in the resected specimen. This margin detection task is currently performed using histological techniques (Cendán et al. 2005) that under-sample the surface of the tumor (Laughney et al. 2012; Carter 1986). As a result, in cases of 20% to 70% of lumpectomy patients, the margin detection gives a false negative result and the patient must return to the operating room for a repeat surgery (Jacobs 2008). XRD imaging offers the possibility to improve the treatment outcomes for these patients due to its ability to spatially resolve and classify cancerous breast tissue.

Griffiths et al. (2008) used the CSCT architecture to perform the first XRD imaging study for breast cancer, doing so in biopsy-sized breast tissue samples ex vivo, similar to the imaging task that would be involved in margin detection of lumpectomy samples. They were able to see contrast between cancerous and healthy tissue with a spatial resolution of 0.5 mm, roughly corresponding to the size-scale of micro calcifications.

Furthermore, Lakshmanan et al. (2015) showed using Monte Carlo simulations that XRD imaging using the CSCT architecture can accurately classify cancerous voxelsinsideofheterogeneousbreasttissuesamplesresembling thoseremovedduring lumpectomy with an AUC of 0.97. That study employed Monte Carlo simulations based on the Geant4 toolkit (Agostinelli et al. 2003) because its electromagnetic physics has been validated for accuracy through several experiments (Apostolakis et al. 2010; Cirrone et al. 2010; Batic et al. 2012). The angular distribution of X-ray coherent scattering in the simulation was modified to incorporate diffraction-effects from electromagnetic interference using the experimentally measured XRD patterns of the different classes of breast tissue (see Figure 4.2(a)). The breast tissue phantom modeled was based on two components:(1) the healthy tissue distribution in the phantom was based on the segmented distributions of adipose, fibroglandular, and mixtures of the two from high-resolution breast CT images of asymptomatic patients (2) the cancerous tissue in the phantom was based on a mathematical tumor model derived from segmented lesions in high-resolution tomosynthesis images.

The XRD images that were obtained from two of the lumpectomy specimens using the optimized CSCT architecture are shown in two dimensions in Figure 4.3 and in three dimensions in Figure 4.4. The original phantoms are shown in the first column, where blue pixels represent adipose tissue, magenta represents fibroglandular, red represents mixtures of adipose and fibroglandular, and black represents cancerous tissue. The second column shows the acquired XRD images using CSCT at momentum transfer $q = 1.1$ nm^{-1} (which is the momentum transfer value where cancerous tissue and healthy tissue show the greatest scatter intensity contrast, as can be confirmed in Figure 4.2(a)). The final column shows cancerous tissue classification results. The classification is based on a pixel-by-pixel evaluation of the l^2-normor Euclidean distance of the reconstructed pixel intensity in the XRD image across q-space from the expected ground truth scatter intensity of cancerous tissue based on ex vivo measurements (e.g., the plot for cancer in Figure 4.2(a)). Volume renderings of the voxels classified as cancerous are shown in Figure 4.4(b) and (d), juxtaposed with volume renderings of the corresponding ground truth tumors in (a) and (c).

The XRD images acquired of the realistic lumpectomy phantom indicate that XRD has potential for accurate cancer imaging. The authors estimate that the volumetric scans presented would require roughly a 10-minute scan time. An XRD architecture that implements coded apertures to potentially reduce this required scan time is discussed next.

4.3.5 XRD Imaging of Lumpectomy Specimens: Coded Aperture Approach

The ~10-minute scan time that was found to be required for the CSCT architecture can be further reduced using coded apertures that encode depth-information. This coded aperture technique for XRD imaging can enable volumetric imaging without the rotation and repeated exposures required

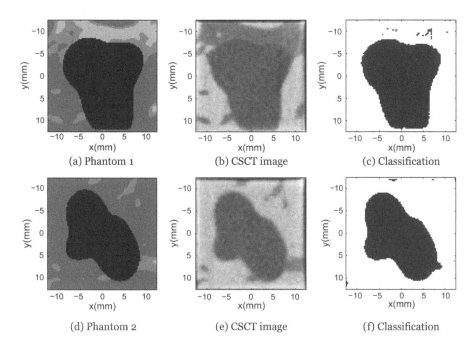

FIGURE 4.3
The two ground truth breast tissue phantoms are shown in the first column (i.e., (a) and (d)), where blue pixels represent adipose tissue, magenta represents fibroglandular, red represents mixtures of adipose and fibroglandular, and black represents cancerous tissue. The second column shows the XRD image acquired; the XRD images are shown for a momentum transfer value of $q = 1.1$ nm^{-1}, which is where cancerous tissue and healthy tissue show the greatest scatter intensity contrast (e.g., see Figure 4.2(a)). The third column shows the results of cancerous tissue classification pixel-by-pixel. The overall cancerous voxel classification accuracy was found to have an AUC of 0.97. ©Institute of Physics and Engineering in Medicine. Reproduced from Lakshmanan et al. (2015) by permission of IOP Publishing. All rights reserved.

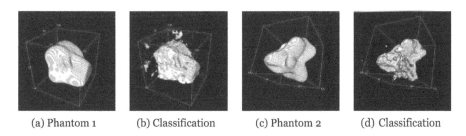

FIGURE 4.4
Three-dimensional volume renderings of the two ground truth breast tumors and the cancerous tissue classifications from Figure 4.3 ©Institute of Physics and Engineering in Medicine. Reproduced from Lakshmanan et al. (2015) by permission of IOP Publishing. All rights reserved.

in computed tomography scan geometries such as in the CSCT architecture. The efficacy of the coded aperture approach is reviewed here.

4.3.5.1 Monte Carlo Simulations

In another study (Lakshmanan et al. 2016) by the same authors who demonstrated the CSCT technique to have an AUC of 0.97, they showed that similar levels of cancerous voxel detection accuracy (92%) could be achieved when using a coded aperture architecture, in which depth resolution is obtained using the coded aperture rather than rotating the source around the sample as in the CSCT design. The study used the same Geant4 Monte Carlo toolkit and realistic high-resolution anthropomorphic breast tissue phantom as from the study using the CSCT architecture discussed in the previous subsection. The imaging system from that study was modified by placing a 1-mm slab of bismuth-tin alloy machined into a series of uniform slits with spatial frequency $0.5\,mm^{-1}$ at 83 mm in front of the detector in order to modulate the scattered X-rays approaching the detector from the sample. In addition, the rotation motion of the sample as is typically done in computed tomography scan acquisitions was eliminated. Instead, the depth information of the sample was reconstructed based on the depth-dependent modulation of the coded aperture by the scattered X-rays. A physical model of the imaging system— that incorporates this depth-dependent modulation and that was developed by Greenberg et al. (2013)—was inversely solved using a maximum *a posteriori* estimation method with total-variation (TV) regularization to reconstruct an image of the object. As in the CSCT study, the Euclidean distance or l^2-norm of the reconstructed scatter intensity across q-space was used to classify voxels as cancerous, adipose, figroglandular, or a mixture of the two.

The resulting coded aperture XRD images are shown in Figure 4.5. The modeled phantom is shown in Figure 4.5(a), where black represents cancerous

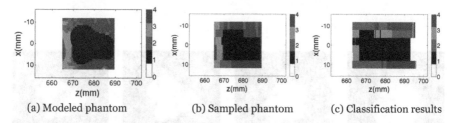

(a) Modeled phantom   (b) Sampled phantom   (c) Classification results

FIGURE 4.5

(a) The modeled phantom, where black represents cancerous pixels, blue represents adipose, magenta represents fibroglandular, and red represents mixtures of adipose and fibroglandular. (b) A version of the phantom subsampled in x based on the 4-mm step size used for the raster scan of the pencil beam in the coded aperture XRD system. This subsampled phantom represents the achievable ground truth object given the raster scan sampling rate. (c) The final reconstructed classified XRD image. This resulting classification has an accuracy for positive (i.e., cancerous) voxel detection of 92%. Reproduced from Lakshmanan et al. (2016) by permission of SPIE. All rights reserved. ©SPIE.

pixels, blue represents adipose, magenta represents fibroglandular, and red represents mixtures of adipose and fibroglandular. Because the beam was raster scanned across x in 4-mm steps, that sampling rate was used to display a subsampled version of the phantom in Figure 4.5(b); this subsampled version of the phantom represents the achievable ground truth object given the sampling rate used. Finally, the reconstructed classified result is shown in Figure 4.5(c). The accuracy of cancerous voxel detection in Figure 4.5(c) for correctly identifying the cancerous (black) pixels in (b) is computed to be 92%.

The accuracy for cancerous voxel detection provided by this coded aperture system is comparable to that obtained using the CSCT architecture. The main improvement of the coded aperture XRD imaging system is that its scan time is reduced from that of the CSCT architecture by a factor equal to the number of projection angles required in the CSCT architecture scan, which was 20 angles for the study presented in the previous subsection. Consequently, the coded aperture XRD system can provide a three-dimensional scan time of less than a minute, down from 10 minutes using the CSCT architecture. A reduction in scan time from 10 minutes to less than a minute is significant because it makes XRD more feasible for clinical applications that often require the patient to remain stationary in uncomfortable positions or maintain breath-holds for the duration of the scan, and for which scan time limits the achievable patient-throughput for clinicians and providers.

4.3.5.2 Experimental Implementation

Due to the shorter required scan time and fewer projections that must be acquired in the coded aperture architecture, it has been implemented experimentally with promising qualitative results that indicate its accuracy for cancer detection. Lakshmanan et al. (2016) implemented the coded aperture architecture based on the previously discussed Monte Carlo simulation model and scanned two surgically excised breast samples. One of the samples (Figure 4.6(a)) was shown to contain a cancerous tumor using routine clinical workup through H&E staining (Cendán et al. 2005), whereas the other was a matched healthy sample (Figure 4.8(a)) from the same patient. The coded aperture system comprised an X-ray tube (Varian model G1593BI) operating at 125 kVp and 50 mA, pinhole collimators, a bismuth-tin alloy coded aperture with $0.5\,\text{mm}^{-1}$ spatial frequency, and an array of 128 energy-sensitive CdTe detector pixels (Multix ME-100 Version2). The detector pixels had size 0.8 mm by 0.8 mm and measured X-ray energies in the approximate range of 20–160 keV using 64 energy bins spaced 2.2 keV apart with FWHM energy resolution of 6.5 keV.

For the cancerous breast sample, Figure 4.6(b) shows a volume rendering of the scatter intensity (i.e., the reconstructed three-dimensional image of scatter intensity integrated over q-space), Figure 4.6(c) shows a single slice

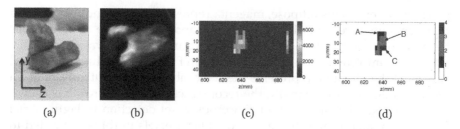

(a)   (b)   (c)   (d)

FIGURE 4.6
For the cancerous surgically excised breast tissue sample scanned using the coded aperture
XRD architecture in experiment: (a) photograph of sample, (b) volume rendering of the reconstructed scatter intensity, (c) a slice through the scatter intensity volume, and (d) the tissue-
classification results (black: cancer, blue: adipose, magenta: fibroglandular, red: mixture of
adipose and fibroglandular). The black pixels in (d) confirm the indications from histology that
the sample contains a cancerous tumor. Reproduced from Lakshmanan et al. (2016) by permission of SPIE. All rights reserved. ©SPIE.

through volume rendering, and Figure 4.6(d) shows the final classification
results. The black pixels (labeled as 'A') indicate the presence of cancer in the
sample, confirming the histology findings that the sample contains a cancerous tumor. Furthermore, the tumor within the cancerous sample is rendered
in three dimensions in Figure 4.7, alongside renderings for the three other
classes of breast tissue. Similar results for the healthy sample are shown in
Figure 4.8. The absence of black pixels in Figure 4.8(c) confirms the histology
findings that the sample is healthy.

The experimental implementation of the coded aperture architecture
shows that it can be used to generate volumetric maps of the distributions of
healthy and cancerous tissue inside of breast samples to potentially replace
histology analysis for assessing tissue samples. Specifically, the experimental implementation demonstrates that the energy-sensitivity of available
detector technology is sufficient and that background scatter from experimental components can be sufficiently suppressed to generate XRD images
for breast tissue classification. To further validate XRD imaging for cancerous breast tissue imaging, a quantitative evaluation of the spatial accuracy
of cancerous voxel detection benchmarked against histology is necessary for

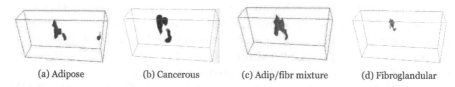

(a) Adipose   (b) Cancerous   (c) Adip/fibr mixture   (d) Fibroglandular

FIGURE 4.7
Volume renderings of the pixels classified as being one of the four tissue categories for the
cancerous sample shown in Figure 4.6 based on experimental results using the coded aperture architecture. Reproduced from Lakshmanan et al. (2016) by permission of SPIE. All rights
reserved. ©SPIE.

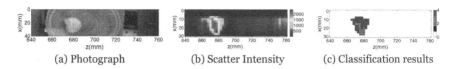

(a) Photograph   (b) Scatter Intensity   (c) Classification results

FIGURE 4.8
For the healthy tissue sample scanned using the experimental implementation of the coded aperture architecture: (a) photograph of the sample, (b) the reconstructed scatter intensity for a single slice, and (c) the tissue-classification results. The absence of the black pixels indicates an absence of cancer, confirming the histology findings that the sample comprises healthy tissue. Reproduced from Lakshmanan et al. (2016) by permission of SPIE. All rights reserved. ©SPIE.

the experimental implementation of XRD breast imaging as has been demonstrated and discussed for the Monte Carlo simulations.

The potential for extending XRD imaging from tissue specimens (~2.5 cm size) to in vivo whole breasts (11 cm median diameter (Boone et al. 2001)) for applications in breast cancer screening and diagnosis mainly involves overcoming the obstacle posed by multiple scatter noise, which is discussed in detail in Section 9.1. However, XRD imaging is likely able to provide high levels of accuracy for breasts that are compressed as in mammography, which have a median thickness of 5.2 cm (Boone et al. 2001). That thickness is well under the ~10 cm size range determined in Section 9.1 to be where multiple scatter noise becomes significant.

4.4 Informing Treatment for Urinary (Kidney) Stones

Urinary (kidney) stones are hard masses composed of minerals that form in the urinary tract, typically causing patients considerable pain in the area between the ribs and hips in the back, as well as leading to bleeding, infection, or the blockage of the flow of urine (Stamatelou et al. 2003). Nearly 19% of men in the United States aged over 70 report have a history of kidney stones. When a man does experience pain and is treated for kidney stones, the likelihood of a recurrence is 80% within 10 years (Scales et al. 2012). Due to its prevalence and associated affliction, available improvements in kidney stone diagnosis and treatment would have considerable applications.

4.4.1 Standard Treatment

Kidney stones can potentially be treated using a variety of options: surgery, other invasive techniques, or non-invasively using extracorporeal shock wave lithotripsy (ESWL) (Chaussy et al. 1984, Perez Castro et al. 2014). Whenever it is available, non-invasive treatment is the preferred approach due to the reduced recovery time for the patient that accompanies the procedure.

However, kidney stones that have certain mineral compositions (urate, calcium oxalate monohydrate or whewellite, and cystine) are unresponsive to
treatment using ESWL (Newman et al. 1987). Therefore, the identification
of urinary stone composition in a patient can inform the efficient planning
of the subsequent course of treatment. In addition to better informing the
choice of treatment, knowledge of the type of kidney stone can help clinicians with planning follow up schedules and strategies for prevention of
kidney stone recurrence (Batchelar et al. 2002).

4.4.2 Potential for XRD Imaging

A number of methods are being researched for in vivo urinary stone composition determination such as urinary pH, urine microscopy, radiography (Ramakumar et al. 1999; Motley et al. 2001), conventional CT (Nakada
et al. 2000), dual-energy CT (Duan et al. 2015; Li et al. 2013a, b), and MRI
(Dawson et al. 1994) with some success, but none of these methods have
been able to provide sufficient material identification or discrimination
ability to detect the broad array of types of kidney stone compositions.
Dual-energy CT, for example, is limited to decomposing samples using two
or at most three basis materials (Liu et al. 2009) because data is available for
only two energy channels, whereas there are seven basis materials whose
presence must be estimated in a urinary stone. Due to its material discrimination ability and the numerous momentum transfer value samples measured (e.g., much beyond the two energies available in dual-energy CT),
XRD imaging maybe well-suited to the task of kidney stone composition
identification. Furthermore, spatially resolved XRD information provided
by XRD imaging would provide more diagnostic information for urolo-
gists by allowing them to better understand the cases or mechanisms for
kidney stone formation and characterizing stone fragmentation by ESWL
(Williams et al. 2003). For example, different kidney stone compositions
show different degrees of fragmentation success from ESWL treatment
(Williams et al. 2003).

The first XRD study of kidney stones was carried out by Dawson et al.
(1996), who were able to arrive at preliminary results demonstrating that
different kidney stones showed different XRD patterns. That work was
expanded when Batchelar et al. (2002) showed through analysis of surgically removed urinary stones that each of the seven types of stone components (calcium oxalate monohydrate, calcium phosphate, calcium phosphate
dihydrate, cystine, magnesium ammonium phosphate, uric acid, calcium
oxalate dihydrate) has a distinct XRD pattern. They concluded that XRD
analysis can be used to decompose urinary stones into the seven basic
stone components due to the differences in XRD patterns exhibited by each
type of stone or basis material. In addition, given the penetrating capability of X-rays, XRD imaging can potentially be translated from in vitro to in
vivo use.

4.4.3 XRD Imaging Feasibility

Davidson et al. (2005a) were able to reconstruct XRD images using the CSCT architecture of a kidney stone phantom that contained inserts of all seven types of kidney stones. They were able to decompose the XRD images into material maps using a non-negative least squares regression. In the material maps, each of the seven types of kidney stones could be identified and isolated from each other. Moreover, they were able to also decompose mixtures of the two types of stones.

The authors extended their XRD imaging implementation to two intact kidney stones samples that had been surgically removed from patients in Davidson et al. (2005b). They were able to show that XRD imaging improves upon infrared spectroscopy (IRS)—the standard clinical laboratory technique for assessing kidney stones. For example, IRS analysis showed that the first kidney stone was solely composed of uric acid, and that the second kidney stone was purely composed of calcium phosphate dihydrate. The material maps determined from the XRD images of the first sample agreed with the IRS assessment. The XRD material maps for the second stone showed that the stone was primarily composed of calcium phosphate dihydrate as indicated by IRS but that smaller regions within the sample contained two other kidney stone components (see Figure 4.9). Therefore, the increased spatial sampling rate over what is available in IRS that is available in XRD imaging was able to reveal additional kidney stone components. Such unrecognized kidney stone components may be significant if they are of the type that are resistant to ESWL treatment and if that therapeutic path was chosen. In that case, the ESWL treatment would prove to have insufficient effectiveness in breaking the stone down to small enough fragments.

4.5 Brain Imaging

Johns et al. (2002) showed using a semi-analytic model that XRD measurements provide better signal-to-noise-ratio (SNR) for distinguishing white and gray matter in the brain than conventional CT for the same radiation dose imparted to the patient. Based on differences in their attenuation coefficients, conventional CT is able to provide a maximum contrast between the two types of brain matter of only 0.5%. For their XRD imaging approach, in order to obtain the spatial sensitivity for X-ray detection necessary to measure scattering angle, they selectively measured X-rays collected in a single ring or annulus at the detector. Despite the reduced signal from this selective detection of scattered x-rays and the accompanying increased statistical noise, the XRD imaging measurements still gave better SNR and contrast

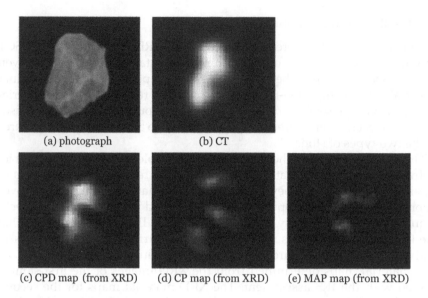

(a) photograph   (b) CT

(c) CPD map (from XRD)   (d) CP map (from XRD)   (e) MAP map (from XRD)

FIGURE 4.9
Subfigures (a) and (b) show a photograph and attenuation map for the second sample studied by Davidson et al. (2005b). Subfigures (c) through (e) show the classified material maps determined from the XRD images for CPD (calcium phosphate dihydrate), CP (calcium phosphate), and MAP (magnesium ammonium phosphate). IRS analysis—the standard clinical lab technique for kidney stone analysis—classified the sample as CPD. However, as shown in the material maps in this figure, XRD imaging was able to reveal a more detailed composition of the sample by detecting the presence of CP and MAP. © Institute of Physics and Engineering in Medicine. Reproduced from Davidson et al. (2005b) by permission of IOP Publishing. All rights reserved.

than conventional CT as long as the target size was less than 40 mm, which is likely due to increased multiple scatter noise from extended targets. The authors were also able to conclude that XRD imaging alone provides better SNR than when also incorporating back-scatter (i.e., Compton scatter) measurements.

The improved brain tissue contrast predicted by this semi-analytic model was demonstrated experimentally by Jensen et al. (2011). They developed a monochromatic (synchrotron-based) XRD imaging system using the CSCT architecture for applications in brain imaging and applied it to a series of rat brains, including one with a cancerous tumor. In order to image the brain, their XRD imaging system was developed to visualize structures on the nanometer size-scale by using a 25 μm focal spot size, 25 μm beam step size, a 0.67°angular sampling rate, a single photon-counting detector and a 4.5-hour scan time per slice. They were able to show that XRD imaging is able to provide anatomical as well as functional information about the brain at this nanometer size-scale, with greater contrast visible than is present in conventional CT (see Figure 4.10).

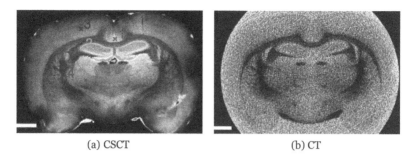

(a) CSCT   (b) CT

FIGURE 4.10
Coronal slices of a healthy rat brain using (a) XRD imaging (CSCT architecture) and (b) conventional CT. The XRD images provide greater contrast between structures than conventional CT due to differences in the XRD signal between grey and white matter. The labeled regions in (a) correspond to regions whose XRD patterns in momentum transfer (q) space are shown in Jensen et al. (2011). The hippocampus structure—the component of the brain responsible for memory (Tulving & Markowitsch 1998)—is marked using the green label "$_x2$". ©Institute of Physics and Engineering in Medicine. Reproduced from Jensen et al. (2011) by permission of IOP Publishing. All rights reserved.

4.6 Liver Imaging

In the clinical workflow for performing liver transplants, a critical step is screening livers that have excessive (>30%) fat content because they pose a risk of postoperative mortality: specifically, there is an accompanying risk of postoperative mortality of 2% for non-fatty livers versus14% for fatty livers (Selzner & Clavien 2000). Furthermore, increased liver fat content also indicates a risk for liver carcinoma (McCormack et al. 2011). Standard clinical techniques for fatty liver screening include visual inspection and palpation, ultrasound, and histological assessment, which have all been shown to be inconsistent and to have significant variability in results (Rey et al. 2009; McCormack et al. 2011) due to inter- and intra-observer variability. Moreover, histological assessment involves invasive biopsy sampling of the patient's liver. Elsharkawy & Elshemey (2013) carried out an ex vivo study where they showed that XRD imaging can quantify liver fat content. They were able to do so by exploiting the linear relationship between fat content and its XRD signal characteristics measured in the liver. For example, the fat content in a given liver sample can be determined from the liver XRD signal intensity at the momentum transfer value corresponding to fat tissue (i.e., at $q = 1.1$ nm^{-1} based on Figure 4.1(a)) as well as the integrated intensity of that peak. The fat peak height and intensity observed in the XRD measurements of the liver samples were found to correlate with the liver fat composition with $R^2>0.9$ for linear dependence and $r>0.9$ for the Pearson's correlation coefficient.

Theodorakou & Farquharson (2009) were also able to exploit the ability to quantify fat content using XRD imaging and were able to apply fat quantification to identify cancerous liver samples. Their multivariate classifier was able to use the peak heights in the XRD signals to identify normal liver tissue samples with 67% accuracy, and secondary colorectal liver cancer tissue samples with 60% accuracy.

4.7 Blood Vessel Plaque Detection

The buildup of plaque in the blood vessels—often called atheroma when occurring in the arteries—can lead to rupture, thrombosis, heart attack, and atherosclerosis, which is the thickening of the artery walls due to white blood cell accumulation and is a leading cause of stroke (Fuster 1999). Despite the serious consequences associated with blood vessel plaque buildup, no diagnostic techniques are currently available that can fully characterize plaque composition in atherosclerosis: MR,CT, and ultrasound can potentially assess and identify plaque size and morphology, but cannot fully determine composition (Pasterkamp et al. 2000; Helft et al. 2001; Estes et al. 1998), which is considered to be the most consistent indicator of disease risk (Fuster 1999).

Davidson et al. (2002) used XRD imaging to produce material-specific images for a surgically excised atherosclerotic (diseased) carotid artery, identifying regions of undiseased vessel tissue, fat, and hydroxyapatite plaque from calcified deposits. They validated their XRD imaging material identification method using a proof-of-concept phantom. Their XRD imaging system used a diagnostic X-ray source and therefore could also be used for in vivo imaging of blood vessels. The challenges that must be addressed for in vivo imaging, however, include multiple scatter noise with increasing penetration, especially for large patients.

4.8 Radiation Dose Analysis

X-rays are a form of ionizing radiation, which in biological tissue cause DNA strand breaks that can induce cancer through mutations, chromosomal translocations, and gene fusions (Brenner & Hall 2007). Increased radiation dose is necessary, however, to obtain sufficient signal- or contrast-to-noise ratio for the diagnostic imaging task at hand. Therefore, an evaluation of the radiation doses required for effective XRD imaging in comparison to standard diagnostic X-ray and CT imaging techniques is a significant factor in assessing its applicability for medical imaging.

Table 4.2 shows the equivalent dose estimates for XRD imaging from various studies, and for comparison, also includes the typical dose delivered

TABLE 4.2

Radiation Dose Requirements for Effective XRD Imaging Compared to Other X-ray Imaging Techniques

XRD Application	Eq. Dose	Study
Bone ex vivo (osteoporosis)	100 mSv	Westmore et al. (1997)
	25 mSv	Batchelar & Cunningham (2002)
Whole breast ex vivo	5.6 mSv	Ghammraoui et al. (2015)
Liver in vivo	0.32–3.2 mSv	Speller et al. (2015)
Fat/muscle/bone contrast	7 mSv	Kleuker et al. (1998)
Other Imaging Technique	*Eq. Dose*	*Reference*
QCT (osteoporosis)	6–300 mSv	Damilakis et al. (2010)
DEXA (osteoporosis)	<0.015 mSv	
Breast CT	2.5–10.3 mSv	Lindfors et al. (2008)
Mammography	3 mSv	Hall & Brenner (2008)
Abdominal CT (liver)	10 mSv	

The data show that XRD imaging involves comparable doses to those required of other X-ray imaging techniques.

from other X-ray imaging techniques. We can see that the dose estimates for XRD imaging are comparable to the typical doses used for other X-ray imaging techniques. Furthermore, there is considerably greater potential for dose reduction in XRD imaging because thus far research into XRD imaging has been primarily focused on demonstrating proof-of-concept and efficacy; whereas dose reduction techniques in mammography and CT is an established line of research that has been pursued for over 40 years (Gordon 1976).

4.9 Current Limitations of XRD Imaging for Medical Applications

4.9.1 Multiple Scatter Noise

X-rays that scatter multiple times have angular and energy characteristics that are stochastic in nature: the probability that an X-ray has two or more scatter events with certain energy transfers and angles is negligible, as multiple scatter events experienced by a quanta are uncorrelated in nature. However, taken together or integrated overall possible outcomes, multiple scatter events accumulate and contribute considerable signal to XRD measurements: for example, Leclair & Johns (1998) used a semi-analytic model to show that multiple scatter from an object of size 20 cm will result in 20% of the measured signal being multiple scatter noise when using a 60 keV X-ray source. (The proportion of multiple to coherent scatter events increases at greater X-ray energies due to the diminishing coherent scatter cross section.) Furthermore, due to the stochastic nature of multiple scattering, its measurement cannot be used to inversely reconstruct insightful properties of the sample. Consequently, as many of the XRD imaging studies reviewed here acknowledge, multiple scatter effects can cause reconstruction artifacts that distort the reconstructed XRD images.

As a result, the studies carried out for medical applications of XRD imaging thus far have primarily been applied for the analysis of relatively small objects (~3 half-value layers for X-rays in the diagnostic energy range, or ~10 cm). Otherwise, larger size objects may introduce multiple scatter noise to a degree that it degrades image quality. For a number of applications such as imaging extremities—for example, the bones within the arms and legs—and compressed breasts, the target size is within this 10 cm size range and therefore multiple scatter noise is less of a concern. However, when translating XRD imaging from analyzing ex vivo kidney stones to imaging urinary stones that are embedded in the kidney of a large patient or from analyzing a rat brain to the much larger human brain, its efficacy can be limited by multiple scatter noise.

A technique proposed for correcting for multiple scatter is to estimate its magnitude using calibration measurements or Monte Carlo simulations and then subtracting that estimated multiple scatter component from all projection measurements and profiles prior to image reconstruction (Harding & Harding 2007). Because multiple scatter noise is stochastic, its magnitude is position-independent, which greatly simplifies the task of estimating its magnitude.

4.9.2 Scan Time

Medical applications of XRD imaging are also limited by scan time. The most common XRD imaging system for medical applications employs a pencil-beam CSCT architecture, which has similar scan speeds to the first-generation of X-ray CT: a pencil beam of X-rays is raster scanned across the sample and then repeated for hundreds of projection angles. XRD imaging using the CSCT architecture has been shown to require volumetric scan-times of 4.5 hours (Jensen et al. 2011). Requiring a patent to remain static for such a period of time would be impractical. Moreover, for ex vivo applications, it is also impractical to require clinicians who would be using the XRD imaging results in their decision-making to wait for such a period of time when much faster alternative standard technologies such as CT are readily available.

Innovative architectures have recently been introduced in hopes of reducing the XRD scan time required. Lakshmanan et al. (2016) applied to ex vivo breast cancer imaging a previously developed coded aperture architecture (Greenberg et al. 2013) with a cancer detection accuracy of 92% and a scan time of 12 minutes. Additionally, recent work by Hassan et al. (2016) presents a fan beam coded aperture architecture that they demonstrate to provide a speedup of 100 times over the pencil beam coded aperture architecture used by Lakshmanan et al. (2016), when using the same components. The fan beam coded aperture architecture could therefore provide volumetric scans for ex vivo breast imaging within 8 seconds. Further efforts should be devoted to applying fan beam coded aperture XRD imaging for medical applications.

4.10 Summary

Through reconstructing the molecular structure of crystallized samples, XRD has revolutionized the field of biochemistry, e.g., revealing the double-helix structure of DNA. However, over the last three decades, spatially resolved XRD imaging has been demonstrated for reconstructing material maps in non-crystallized samples, such as human soft tissue in vivo, providing potential diagnostic value for numerous clinical scenarios.

XRD imaging has been shown to outperform the standard diagnostic techniques—dual energy X-ray absorptiometry (DEXA) and quantitative CT (QCT)—for mapping the strength properties of the bone correlated with the risk of bone fracture that accompanies the onset of osteoporosis. For breast cancer diagnosis, XRD imaging overcomes the limitation of transmission-based imaging techniques such as mammography and breast CT of minimal and undetectable contrast between fibroglandular and cancerous tissues in the breast by providing consistent and reproducible contrast for cancerous tissue. Similar to the case of breast imaging, XRD imaging is able to provide considerably greater soft tissue contrast than possible in conventional CT, specifically between white and grey matter, enabling clearer visualization of brain structures and tumors.

Due to the high dimensionality provided by the many momentum transfer sample points reconstructed in XRD imaging, it is able to decompose kidney stones into the sufficient number of basis materials to detect the presence of the seven types of components of a kidney stone, which is beyond the material decomposition capability of the other pursued imaging techniques. This material decomposition ability provided by XRD imaging can guide clinicians in choosing the most efficient course of treatment for kidney stone patients. As in the case of kidney stones, for blood vessel plaque, XRD imaging can provide greater material decomposition ability for plaque detection than obtainable using conventional imaging techniques such as MR and CT. Due to the distinct XRD peak for fat, XRD imaging has been shown to be effective in estimating the fat content of livers, which can spare organ donors of invasive biopsy procedures when being screened for liver transplants.

Although the doses required in XRD imaging are comparable to those in other X-ray imaging techniques, two limitations that exist are multiple scatter noise for targets larger ~10 cm and excessively long scan times (~4 hours). Multiple scatter noise can be corrected by estimating its magnitude using calibration and Monte Carlo simulations, followed by scatter subtraction prior to reconstruction. Recent studies have shown that fan beam coded aperture architectures for XRD imaging can potentially provide scan times of less than 8 seconds for ex vivo tissue samples and should therefore be pursued further for medical applications.

References

Agostinelli S, Allison J, Amako K, Apostolakis J, Araujo H, Arce P et al. 2003 *Nuclear Instruments and Methods A* 506(3), 250–303.

Apostolakis J, Bagulya A, Elles S, Ivanchenko V, Jacquemier J, Maire M, Toshito T & Urban L 2010 *Journal of Physics: Conference Series* 219(3), 032044.

Batchelar D L, Chun S S, Wollin T A, Tan J K, Beiko D T, Cunningham I A & Denstedt J D 2002 *The Journal of urology* 168(1), 260–5.

Batchelar D L & Cunningham I A 2002 *Medical Physics* 29(8), 1651.

Batchelar D L, Davidson M T M, Dabrowski W & Cunningham I A 2006 *Medical Physics* 33(4), 904.

Batic M, Hoff G, Pia M G & Saracco P 2012 *IEEE Transactions on Nuclear Science* 59(4), 1636–1664.

Bear R S 1942 *Journal of the American Chemical Society* 64(3), 727–727.

Blundell T, Dodson G, Hodgkin D & Mercola D 1972 *Advances in Protein Chemistry* 26, 279–402.

Bonucci E 1998 in *'Bone Densitometry and Osteoporosis'* Springer pp. 173–191.

Boone J M, Nelson T R, Lindfors K K & Seibert J A 2001 *Radiology* 221(3), 657–667.

Bragg W & Bragg W 1913 *Proceedings of the Royal Society of London. Series A* 88(605), 428–438.

Brenner D J & Hall E J 2007 *New England Journal of Medicine* 357(22), 2277–2284.

Carlström D & Finean J 1954 *Biochimica et biophysica acta* 13, 183–191.

Carter D 1986 *Human Pathology* 17(4), 330–332.

Castro C R F, Barroso R C & Lopes R T 2005 *X Ray Spectrometry* 34(6), 477–480.

Cendán J C, Coco D & Copeland E M 2005 *Journal of the American College of Surgeons* 201(2), 194–198.

Chan H P, Vyborny C J, Macmahon H, Metz C E, Doi K & Sickles E A 1987 *Investigative Radiology* 22(7), 581–589.

Changizi V, Kheradmand A A & Oghabian M A 2008 *Journal of Medical Physics / Association of Medical Physicists of India* 33(1), 19–23.

Changizi V, Oghabian M A, Speller R, Sarkar S & Kheradmand A A 2005 *International Journal of Medical Sciences* 2(3), 118–121.

Chaussy C, Schmiedt E, Jocham D, Schuller J, Brandl H & Liedl B 1984 *Urology* 23(5), 59–66.

Cirrone G, Cuttone G, Di Rosa F, Pandola L, Romano F & Zhang Q 2010 *Nuclear Instruments and Methods A* 618(1), 315–322.

Conceiçáco A L C, Antoniassi M & Poletti M E 2009 *Analyst* 134(6), 1077–1082.

Cunha D M, Oliveira O R, Pérez C A & Poletti M E 2006 *X Ray Spectrometry* 35(6), 370–374.

Batchelar D L, Dabrowski1 W & Cunningham I A 2000 *Proceedings of Society of Photographic Instrumentation Engineers (SPIE)* 3977, 353–361.

Damilakis J, Adams J E, Guglielmi G & Link T M 2010 *European Radiology* 20(11), 2707–2714.

Davidson M T M, Batchelar D L & Cunningham I A 2002 in *'SPIE Medical Imaging'* pp. 704–712.

Davidson M T M, Batchelar D L, Velupillai S, Denstedt J D & Cunningham I A 2005a *Physics in Medicine and Biology* 50(16), 3773–3786.

Davidson M T M, Batchelar D L, Velupillai S, Denstedt J D & Cunningham I A 2005b *Physics in Medicine and Biology* 50(16), 3907–3925.

Dawson C, Aitken K, Ng Y, Dolke G, Gadian D & Whitfield H 1994 *Urological Research* 22(4), 209–212.

Dawson C, Horrocks J A, Kwong R, Speller R D & Whitfield H N 1996 *World Journal of Urology* 14 Suppl 1(0724–4983 (Print)), S43–S47.

Debye P & Scherrer P 1916 *Nachrichten von der Gesellschaft der Wissenschaften zu Göttingen, Mathematisch-Physikalische Klasse* 1916, 1–15.

Deisenhofer J, Epp O, Miki K, Huber R & Michel H 1985 *Nature* 318, 618–624.

Desantis C E, Fedewa S A, Sauer A G, Kramer J L, Smith R A & Jemal A 2016 66(1), 31–42.

Duan X, Li Z, Yu L, Leng S, Halaweish A F, Fletcher J G & Mccollough C H 2015 *American Journal of Roentgenology* 6(205), 1203–1207.

Elsharkawy W B & Elshemey W M 2013 *Radiation Physics and Chemistry* 92, 14–21. doi:10.1016/j.radphyschem.2013.07.010.

Elshemey W M, Desouky O S, Fekry M M, Talaat S M & Elsayed A a 2010 *Medical Physics* 37(8), 4257.

Elshemey W M, Mohamed F S & Khater I M 2013 *Radiation Physics and Chemistry* 90, 67–72.

Estes J, Quist W, LoGerfo F & Costello P 1998 *Journal of Cardiovascular Surgery* 39(5), 527.

Evans S H, Bradley D A, Dance D R, Bateman J E & Jones C H 1991 *Physics in Medicine and Biology* 36(1), 7–18.

Friedrich W, Knipping P & Laue M v 1913 *Annals of Physics (Berlin)* 41, 971–988.

Fuster V 1999 *Cardiovascular Drugs and Therapy* 13(4), 363–363.

Ghammraoui B & Badal A 2014 *Physics in medicine and biology* 59(13), 3501–3516.

Ghammraoui B, Badal A & Popescu L M 2016 *Physics in Medicine and Biology* 61(8), 3164–3179. URL: http://stacks.iop.org/0031-9155/61/i=8/a=3164?key=crossref.c587e9e7e1a2afa5c49f2d4c22590176.

Ghammraoui B, Popescu L M & Badal A 2015 in *'SPIE Medical Imaging'* pp. 94121G–94121G.

Gordon R 1976 *Investigative Radiology* 11(6), 508–517.

Grampp S, Genant H K, Mathur A, Lang P, Jergas M, Takada M, Glüer C C, Lu Y & Chavez M 1997 *Journal of Bone and Mineral Research* 12(5), 697–711.

Greenberg J A, Krishnamurthy K & Brady D 2013 *Optics Express* 21(21).

Griffiths J, Royle G, Horrocks J, Hanby A, Pani S & Speller R 2008 *Radiation Physics and Chemistry* 77(4), 373–380.

Gülsün M, Demirkazik F B & Ariyürek M 2003 *European Journal of Radiology* 47(3), 227–231.

Hall E J & Brenner D J 2008 *British Journal of Radiology* 81(965), 362–378.

Hall F M, Davis M A & Baran D T 1987 *New England Journal of Medicine* 316(4), 212–214.

Harding G & Harding A 2007 in 'Counterterrorist Detection Techniques of Explosives' Elsevier pp. 119–235.

Harding G, Kosanetzky J & Neitzel U 1987 *Medical Physics* 14(4), 515–525.

Hassan M, Greenberg J A, Odinaka I & Brady D J 2016 *Optics Express* 24(16), 111–114.

Helft G, Worthley S G, Fuster V, Zaman A G, Schechter C, Osende J I, Rodriguez O J, Fayad Z A, Fallon J T & Badimon J J 2001 *Journal of the American College of Cardiology* 37(4), 1149–1154.

Hull A W 1917 *Physical Review* 10(6), 661.

Jacobs L 2008 *Annals of Surgical Oncology* 15(5), 1271–1272.

Jacobsen K K, O'Meara E S, Key D, Buist D S M, Kerlikowske K, Vejborg I, Sprague B L, Lynge E & Von Euler-Chelpin M 2015 *International Journal of Cancer* 137(9), 2198–2207.

Jensen T H, Bech M, Bunk O, Thomsen M, Menzel A, Bouchet A, Le Duc G, Feidenhans'l R & Pfeiffer F 2011 *Physics in Medicine and Biology* 56(6), 1717–1726.

Johns P C, Leclair R J & Wismayer M P 2002 *SPIE Optics + Optoelectronics* 357(May), 355–357.

Johns P C & Yaffe M J 1987 *Physics in Medicine and Biology* 32(6), 675–695.

Kidane G, Speller R D, Royle G J & Hanby A M 1999 *Physics in Medicine and Biology* 44(7), 1791–802.

Kleuker U, Suortti P, Weyrich W & Spanne P 1998 *Physics in Medicine and Biology* 43(10), 2911–2923.

Kosanetzky J, Knoerr B, Harding G & Neitzel U 1987 *Medical Physics* 14(4), 526.

Laii H & Cunningham I A 1998 3336(May), 707–715.

Lakshmanan M N, Greenberg J A, Samei E & Kapadia A J 2016 *Journal of Medical Imaging* 3(1), 013505.

Lakshmanan M N, Harrawood B P, Samei E & Kapadia A J 2015 *Physics in Medicine and Biology* 60(16), 6355–6370.

Laughney A M, Krishnaswamy V, Rizzo E J, Schwab M C, Barth R J, Pogue B W, Paulsen K D & Wells W a 2012 *Clinical Cancer Research: An Official Journal of the American Association for Cancer Research* 18(22), 6315–6325.

Leclair R J & Johns P C 1998 *Medical Physics* 25(6), 1008–1020. URL: www.ncbi.nlm. nih.gov/pubmed/9650191.

Lewis R a, Rogers K D, Hall C J, Towns-Andrews E, Slawson S, Evans A, Pinder S E, Ellis I O, Boggis C R, Hufton a P & Dance D R 2001 *Proc SPIE* 4320, 547–554.

Li X H, Zhao R, Liu B & Yu Y Q 2013a *Clinical Radiology* 68(7), e370–e377.

Li X, Zhao R, Liu B & Yu Y 2013b *Urology* 81(4), 727–730.

Lindfors K K, Boone J M, Nelson T R, Yang K, Kwan A L C & Miller D F 2008 *Radiology* 246(3), 725–733.

Liu X, Yu L, Primak A N & McCollough C H 2009 *Medical Physics* 36(5), 1602–1609.

McCormack L, Dutkowski P, El-Badry A M & Clavien P A 2011 *Journal of Hepatology* 54(5), 1055–1062.

Motley G, Dalrymple N, Keesling C, Fischer J & Harmon W 2001 *Urology* 58(2), 170–173.

Nakada S Y, Hoff D G, Attai S, Heisey D, Blankenbaker D & Pozniak M 2000 *Urology* 55(6), 816–819.

Newman D, Lingeman J, Mertz J, Mosbaugh P, Steele R & Knapp Jr P 1987 *The Urologic Clinics of North America* 14(1), 63–71.

Newton M, Hukins D W L & Harding G 1992 *Physics in Medicine and Biology* 37(6), 1339–1347.

Oliveira O R, Conceição A L C, Cunha D M, Poletti M E & Pelá C A 2008 *Journal of Radiation Research* 49(5), 527–532.

Pasterkamp G, Falk E, Woutman H & Borst C 2000 *Journal of the American College of Cardiology* 36(1), 13–21.

Perez Castro E, Osther P J S, Jinga V, Razvi H, Stravodimos K G, Parikh K, Kural A R & De La Rosette J J 2014 *European Urology* 66(1), 102–109.

Perutz M, Kendrew J & Watson H 1965 *Journal of Molecular Biology* 13(3), 669–678.

Poletti M E, Goncalves O D & Mazzaro I 2002 *X-Ray Spectrometry* 31(1), 57–61.

Pucci-Minafra I, Andriolo M, Basirico L, Alessandro R, Luparello C, Buccellato C, Garbelli R & Minafra S 1998 *Carcinogenesis* 19(4), 575–584.

Pucci-Minafra I, Luparello C, Andriolo M, Basirico L, Aquino A & Minafra S 1993 *Biochemistry* 32(29), 7421–7427.

Ramakumar S, Patterson D E, LeRoy A J, Bender C E, Erickson S B, Wilson D M & Segura J W 1999 *Journal of Endourology* 13(6), 397–401.

Raymond W A & Leong A S 1991 *Pathology* 23(4), 291–297.

Rey J, Wirges U, Dienes H & Fries J 2009 in 'Transplantation proceedings' Vol. 41 Elsevier pp. 2557–2560.

Rosenbaum D M, Rasmussen S G & Kobilka B K 2009 *Nature* 459(7245), 356–363.

Royle G J, Farquharson M, Speller R & Kidane G 1999 *Bone* 56, 247–258.

Royle G J & Speller R D 1991 *Physics in Medicine & Biology* 36(0031–9155 (Print)), 383–389.

Royle G J & Speller R D 1995 *Physics in Medicine & Biology* 40(0031–9155 (Print)), 1487–1498.

Ryan E A & Farquharson M J 2007 *Physics in Medicine & Biology* 52(22), 6679–6696.

Scales C D, Smith A C, Hanley J M & Saigal C S 2012 *European Urology* 62(1), 160–165.

Schönermark M P, Bock O, Büchner A, Steinmeier R, Benbow U & Lenarz T 1997 *Nature Medicine* 3(10), 1167–1171.

Selzner M & Clavien P A 2000 in '*Seminars in Liver Disease*' Vol. 21 pp. 105–113.

Sidhu S, Falzon G, Hart S a, Fox J G, Lewis R A & Siu K K W 2011 *Physics in Medicine & Biology* 56(21), 6779–6791.

Speller R 1999 *X-Ray Spectrometry* 28(February), 224–250.

Speller R, Abuchi S, Zheng Y, Vassiljev N, Konstantinidis A & Griffiths J 2015 *Journal of Physics: Conference Series* 637, 012026.

Stamatelou K, Francis M, Jones C, Nyberg L & Curhan G 2003 *Kidney International* 63, 1817–1823.

Theodorakou C & Farquharson M J 2009 *Physics in Medicine and Biology* 54(16), 4945–4957.

Tulving E & Markowitsch H J 1998 *Hippocampus* 8(3), 198–204.

Wasnich R D 1996 *Primer on the Metabolic Bone Diseases and Disorders of Mineral Metabolism. Philadelphia, Pa: Lippincott-Raven* pp. 249–251.

Watson J D, Crick F H et al. 1953 *Nature* 171(4356), 737–738.

Westmore M S, Fenster A & Cunningham I a 1997 *Medical Physics* 24(1), 3–10.

Williams J C, Saw K C, Paterson R F, Hatt E K, McAteer J A & Lingeman J E 2003 *Urology* 61(6), 1092–1096.

5

Materials Science of X-Ray Diffraction

Scott D. Wolter

Elon University

CONTENTS

5.1 Introduction

Processing–structure–property (PSP) relationships fundamental to materials science establish that material properties are in large part a consequence of elemental composition and material structure [1]. Accordingly, substances considered as potential threats to human well-being such as explosives have unique structural characteristics that differentiate them from substances considered as non-threats. It is therefore understandable that accessing the structure of objects at the atomic scale in a non-destructive and time-efficient manner would be of interest for security applications. In this chapter, we examine structural aspects of materials as the basis for substance identification in aviation security X-ray diffraction imaging systems.

5.1.1 Materials in Society

The impact and significance of materials breakthroughs in our global society are quite evident considering the rapid rate of technological advances in human history. The transition from the Stone Age to more advanced metallurgical ages was a corollary to our understanding of the diversity of materials in the natural environment and the means to extract and refine these. In recent history, in-depth understanding of the science of materials has enabled processing of specialized metals and alloys, ceramics, semiconductors, and polymers relying on relatively new scientific concepts and materials characterization techniques.

The historical timeline shown in Figure 5.1 reveals important scientific breakthroughs contributing to the science of materials. The Age of Enlightenment was significant as this historical period advanced pursuit of scientific knowledge in the burgeoning disciplines of chemistry and physics. Early during this time period, the existence of chemical elements, the building blocks of matter, was discovered by Robert Boyle based on experimental work fundamentally similar to modern-day chemistry [2]. Subsequently, chemical thermodynamics was founded establishing the relation of heat and work on the state of matter and was profoundly important as it foretold a means to design and engineer materials for specific applications. The atomic level structure of matter was unveiled by such works as W.H. Miller's and A. Bravais' reports of the crystal lattice construct [3,4], the establishment of the complete model of the atom [5,6], W.C. Röntgen's discovery of X-rays [7], and M.T.F. von Laue's discovery of X-ray diffraction [8]. X-rays have not only been used as a scientific tool to probe the elemental composition, crystallinity, and grain structure of materials but have also shown utility in medical applications for imaging and treatment of diseases and, more recently, for use in security applications.

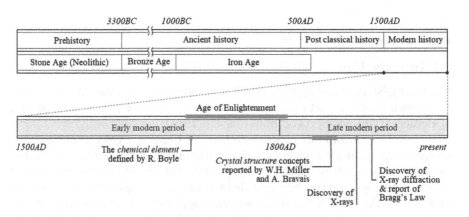

FIGURE 5.1
Timeline of time periods as they relate to major advances in the science of materials.

X-ray diffraction (XRD) leverages well-established scattering phenomena for materials structure determination and, therefore, of interest for airport baggage screening applications. Baggage inspection systems configured to collect scattered X-rays could greatly minimize uncertainty for identification of objects based on transmission imaging alone. Initial work exploiting scattered X-rays for aviation security applications was begun in the 1990s [9–11], but the feasibility of commercializable systems has been hampered by practical detectors, inadequate computer processor technology, and algorithms for spectra data processing. Recent work in our laboratories has explored coded aperture X-ray scatter imaging, which provides a reference structure to ascertain superposed diffraction patterns for object identification [12–14].

Chemical elements that populate the earth have led to the formation of well over 3,000 documented minerals, those that are formed by geological processes. Many more man-made materials—including synthetic plastics, medicines and other healthcare products, structural materials, electronic materials, and so on—have been produced with an increasing number of new materials introduced each year. Commercial products comprised of these collective substances in their pure state or as mixtures could conceivably be found in airport baggage. It is therefore important in X-ray diffraction imaging systems that the measured XRD spectra are scrutinized for reliable substance classification.

5.1.2 Bragg's Law and X-Ray Diffraction

Materials structure determination through examination of XRD patterns has been the focus of intensive effort since von Laue's breakthrough discovery over a century ago [8]. X-rays interacting with matter do so in fundamental ways, each contributing to attenuation of incident photons through the photoelectric effect, Compton effect, coherent scatter, and pair production. X-rays can interact elastically via coherent scatter where, unlike in the Compton effect, there is no energy absorption. The scattering process involves electromagnetic wave interaction with atoms within a material and, consequently, is attuned to the atomic and molecular structure of a material.

The fundamental law to explain coherent scatter in relation to incident X-ray wavelength and atom configuration was reported by father and son physicists, William Henry Bragg and William Lawrence Bragg, that enhanced our understanding of the means of X-ray interactions with matter [15]. The law explains that X-ray diffraction occurs when electromagnetic waves interacting with lattice planes undergo constructive interference. The equation that represents this interaction as a function of scattering angle, Θ, is given as,

$$n\lambda = 2d \sin\Theta$$

whereby n is an integer, d is the interatomic spacing, and λ is the X-ray wavelength. The difference in the incident and diffracted beam angle is given as 2Θ, as shown in Figure 5.2a; often, XRD spectra are plotted as intensity versus Θ–2Θ diffraction angle. The Bragg's equation relates dimensional aspects of a crystalline lattice to angles associated with coherent and incoherent scatter whereby the X-ray wavelength is similar dimensionally to the diffraction plane spacing. Figure 5.2b shows an example of Bragg's law where beam-2 travels $2 \times d \sin \Theta$ longer relative to beam-1. X-rays are constructively scattered when $2 \times d \sin \Theta$ equals $n\lambda$ of the source radiation. Selection rules for diffraction indicate allowed reflections from crystallographic planes that can be used for materials determination. Thus, diffraction spectra may be used as a means of finger print identification for materials possessing long-range structural order. It will be shown as well in Section 5.3 that materials lacking long-range crystallinity, either in solid or liquid form, may also exhibit unique spectral features but is based on a different diffraction modality.

The X-ray diffraction spectra reported herein were acquired using Panalytical X'Pert PRO MRD HR and Bruker D2 PHASER X-ray diffraction systems configured in the Bragg–Brentano geometry. A Cu Kα (8.04 keV at $\lambda = 1.5418\,\text{Å}$) X-ray source was used in both systems, collimated using Soller slits and filtered via nickel foils to remove the Kβ component. Substances were analyzed using Θ–2Θ scans in the range from $5°$–$120°$ at $0.5°$ step size and 0.15–0.5 sec dwell time. Spectra are plotted as a function of momentum transfer (q), which is inversely related to the effective lattice parameter of the material. No correction has been applied to the spectra to account for diffractometer analyzing parameters which may impact relative intensities. It is also noted that the elevated counts at low q in all the reported spectra are caused in part from the primary X-ray beam. This chapter will report on many of the

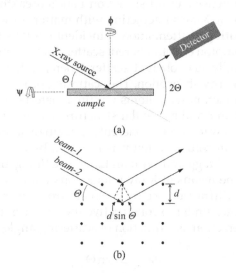

(a)

(b)

FIGURE 5.2
Illustrations of (a) the Bragg–Brentano geometry and angular degrees of freedom and (b) Bragg reflections with the sample viewed as lattice points.

over 400 X-ray spectra of materials acquired for creation of a reference database for coded aperture X-ray scatter imaging system development [12–14].

5.2 X-Ray Diffraction of Crystalline Substances

A commonsensical approach to the discussion of X-ray coherent scatter for differentiating materials type is by considering the degree of structural order within a substance. The extremes in such cases are those materials that are crystalline and those that are either disordered crystalline solids or amorphous substances, exhibiting high to no structural order, respectively. We will begin by considering materials that are crystalline in nature characterized by atoms or molecules configured in a pattern periodic in three dimensions. Spectral attributes of amorphous solids and liquids will be considered in the next section.

5.2.1 Crystallinity and Long-Range Order

Born out of advances in X-ray structural characterization, crystallography is a field of science used to determine atomic arrangements in crystalline solids and forms its geometrical basis through the work of Nicolaus Steno and René-Just Haüy [16]. Steno's law of constancy of angles states that angles between similar crystal facets and associated planes are constant and repeatable throughout crystalline materials. The law of rational indices was deduced by Haüy reporting that crystalline planes intersect crystallographic axes in whole number ratios.

One of the foundational studies applied to the field of crystallography was performed in the mid-19th century by Auguste Bravais and was purely mathematical in origin. Bravais considered symmetric arrays of points to define 14 types of discrete lattices [4]. A three-dimensional array of lattice points is formed at the intersection of lattice planes due to long-range periodicity, as illustrated in Figure 5.3 for the generation of a single lattice point shown at the corner of a cubic lattice. As shown in Figure 5.4, these Bravais lattices are based on seven crystal systems, namely, cubic, tetragonal, orthorhombic, hexagonal, trigonal, monoclinic, and triclinic and four types of unit cells, namely, primitive, body-centered, base-centered, and face-centered unit cells. Primitive lattices are distinct from other forms, in that these contain only one lattice point per unit cell. It is common to distinguish the arrangement of lattice points by their so-called unit cell, defining the most basic structure and symmetry of the lattice. This is an important understanding, in that this gives way to the unique X-ray scattering behavior for crystalline materials. A unit cell is characterized by its physical dimension and shape used to define its lattice parameter or lattice constant.

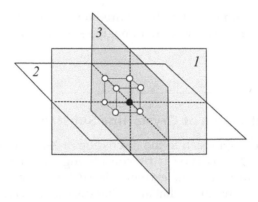

FIGURE 5.3
Illustration of three intersecting lattice planes resulting in the formation of a lattice point, designated by the solid dot on the corner of the cubic cell.

The breakthrough in understanding that atoms and molecules can be considered as residing on lattice sites and follow the configuration of lattice points put forward by Bravais was revolutionary. In this context, atoms or molecules reside on lattice points or fixed positions in relation to lattice points. The spatial arrangement of the lattice planes gives way to peak positions in diffraction spectra based on Bragg's law. Lattice periodicity and X-ray wavelengths on the order of the lattice parameter cause constructive interference with incident X-rays producing Bragg peaks, a means to uncover the lattice parameter, unit cell structure, defects, or otherwise characterize matter based on its crystallinity.

It is understandable that lattice symmetry is pertinent to the discussion of coherent scatter as this is related to the number of unique diffraction planes. For cubic systems, the high degree of symmetry along each of the crystallographic axes gives way to relatively sparse diffraction features. Whereas, for non-cubic systems, less crystalline symmetry leads to more dense diffraction spectra related to a greater number of unique crystalline planes. For instance, the transition from cubic structure to tetragonal and orthorhombic crystal classes causes splitting of a single peak to two peaks for materials with tetragonal symmetry and three peaks for orthorhombic symmetry [17]. This is due to diffraction originating from the same crystalline plane but with two and three different interplanar distances for the respective crystal structures. Each spectral peak is produced from a specific lattice plane as typically reported in XRD spectra using integer values. Nomenclature exists to label these crystallographic planes according to their so-called Miller indices [3] as shown in Figure 5.5 for allowed diffraction planes within the face-centered cubic (fcc) unit cell. The location of the crystallographic planes within this unit cell was determined by the reciprocal of the reported (hkl) planes assuming a lattice parameter of unity.

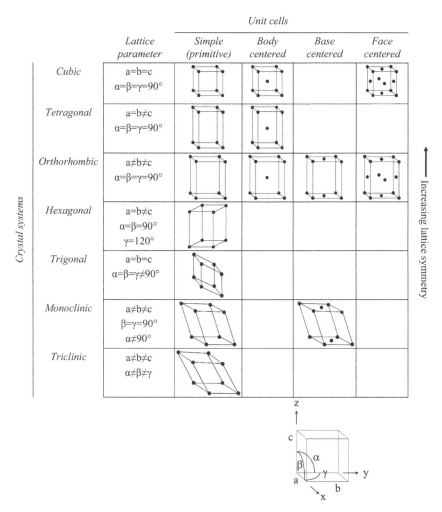

FIGURE 5.4
Bravais lattices and general trend of increasing crystalline symmetry.

5.2.2 Spectra Features of Atomic and Molecular Crystals

It is instructive to consider diffraction features unique to crystalline substances. Figure 5.6 for polycrystalline aluminum powder illustrates the features that characterize a highly crystalline substance possessing a high degree of crystalline symmetry. The narrow width of the peaks is indicative of aluminum atoms on well-defined lattice sites. The corresponding peak positions may be used to determine the lattice parameter of the unit cell and interplanar spacing of the crystalline planes employing Bragg's law. It should be noted that the peak positions and relative peak intensities for many substances may be compared to a database of powder diffraction patterns,

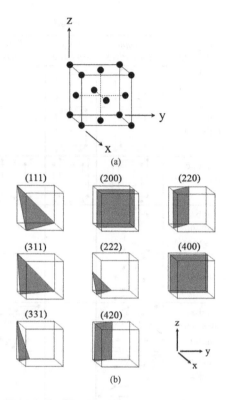

FIGURE 5.5
(a) Depiction of a face-centered cubic unit cell and orthogonal coordinates. (b) Lattice planes associated with allowed reflections for face-centered cubic materials.

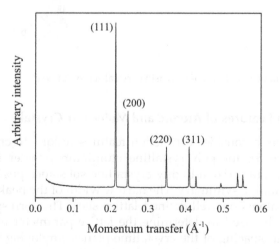

FIGURE 5.6
X-ray diffraction spectrum for aluminum powder.

known as ICDD files (ICDD, International Centre for Diffraction Data) [18]. The characteristic decrease in peak intensity with incident X-ray radiation arises from a decreasing atomic form factor with increasing scattering angle. The spectrum for aluminum represents a random distribution of crystalline powder or grains, as can be confirmed by the narrowness of the diffraction peaks, their peak positions, and relative peak intensities.

The fcc aluminum lattice possesses clearly distinguishable and relatively sparse diffraction features, as expected for cubic materials. A comparison of several fcc elemental metals shows similar diffraction attributes but with diffraction lines at different peak positions. For instance, copper, lead, and aluminum all exist in an fcc configuration; yet, the diffraction spectra are shifted based on differences in their corresponding lattice parameters, as shown in Figure 5.7. Consistent with Bragg's law, the peaks shifted to lower momentum transfer values are due to a somewhat larger lattice for lead (Pb) relative to the other fcc materials.

We next consider a common crystal class but of a different unit cell type, for instance, body-centered cubic (bcc) versus fcc. While the crystal class for both is cubic, the existence of atoms at body-centered and face-centered sites within the two unit cells results in the development of different lattice planes and distinctly different diffraction patterns; the spectral appearance of bcc tungsten, shown in Figure 5.8a, is different from its fcc counterparts. As expected, the decreasing unit cell symmetry of hexagonal closed-packed (hcp) magnesium produces denser and more numerous peaks, as evidenced in Figure 5.8b; yet, it possesses the characteristic diffraction pattern for hcp materials.

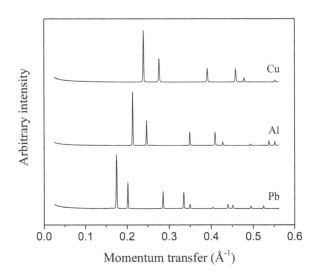

FIGURE 5.7
X-ray diffraction spectra for fcc copper, aluminum, and lead powder. Diffraction patterns look similar but shifted in relation to unit cell size.

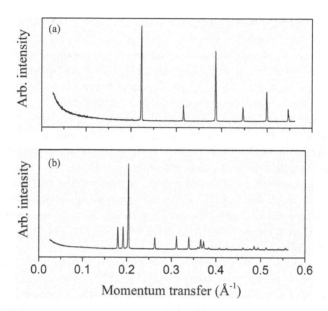

FIGURE 5.8
X-ray diffraction spectra for (a) bcc tungsten and (b) hcp magnesium powder.

The lattice of points proposed by Bravais can be thought of as not only reference points for atoms but also for poly atomic ions and organic molecules. Rather than atoms residing on or near lattice sites, polyatomic ions may configure themselves on or near lattice sites into a Bravais lattice type, such as the $NH_4^+ - NO_3^-$ ion pairing in ammonium nitrate. The diffraction spectrum in Figure 5.9a for ammonium nitrate powder reveals dense and numerous diffraction lines tied to its orthorhombic crystal structure and lower crystalline symmetry in addition to the multiple atoms associated with each lattice site. The interplanar spacing is impacted by the large ionic compound, which results in a larger lattice parameter and spectral features observed at lower momentum transfer values. Molecular crystals, such as sucrose, share a similar lattice structure to polyatomic ionic compounds. In this case, the organic molecules reside on or near lattice sites, which produces diffraction features that are similar to the poly atomic ions, as evidenced in Figure 5.9b for the orthorhombic sucrose powder specimen.

A compilation of spectra showing the peak intensity details (with darker features corresponding to larger peak intensities) are shown in Figure 5.10 for the aforementioned materials discussed in this section. Clearly evident are the distinct patterns that arise for materials with cubic structure based on the geometric and long-range periodicity of the crystalline lattice and are useful for identifying unknown crystalline substances. Diffraction spectra exhibiting characteristic patterns are understandable considering that crystalline materials can be envisioned as point lattices possessing long-range

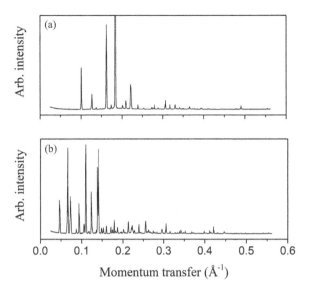

FIGURE 5.9
X-ray diffraction spectra for (a) poly atomic ammonium nitrate and (b) molecular crystalline
sucrose. Both are characterized by orthorhombic crystal structures.

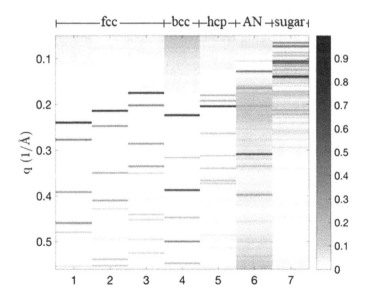

FIGURE 5.10
X-ray diffraction spectra, from left-to-right, for fcc copper, aluminum, lead; bcc tungsten; and
hcp magnesium metal powders contrasted with ammonium nitrate (AN) and sucrose (sugar).
The brightness of the data indicates the spectra intensity, with black and white corresponding
to larger and smaller intensities, respectively.

periodicity [19]. This is highlighted for fcc materials in this figure, which show similar repeatable and periodic patterns among these metals. The bcc materials show single peaks separated at somewhat constant intervals which is a characteristic feature for this crystal structure.

5.2.3 Spectral Anomalies Due to Crystalline Texture

As substance identification in X-ray diffraction imaging systems is reliant upon spectra acquisition and indexing, deviation from expected diffraction patterns is of great concern. For many crystalline materials, spectra variability is a real possibility, a consequence of materials processing that can affect their crystalline state. Consequently, an apparent dependence on sample orientation relative to the incident X-ray beam will arise, showing distinctly different diffraction patterns.

Fiber and/or sheet texture is highly likely for certain materials processing (e.g. roll-formed metals). Sheet texture defines the preferred orientation of crystalline planes normal to the sample surface; while, fiber texture defines the preferred in-plane direction of the crystalline grains. Crystalline texture arising from sheet-forming operations and recrystallization texture from post-annealing have been studied extensively [20–23]. The texture from metal-sheet-forming processes is typically represented as {hkl} <uvw>, where the {hkl} planes are parallel to and the direction <uvw> of the polycrystalline grains are preferentially aligned with the rolling direction [20].

The authors analyzed several roll-formed aluminum samples from different vendors, each exhibiting characteristic attributes of texture, as exemplified in Figure 5.11. The spectral peak positions are consistent with those for fcc aluminum; however, the relative peak intensities for each metal sample

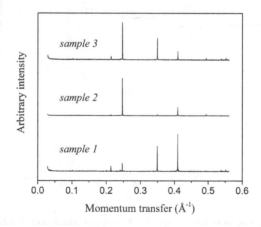

FIGURE 5.11
X-ray diffraction spectra of three samples of roll-formed aluminum sheet from different vendors.

deviates from that of the aluminum powder diffraction specimen shown in Figure 5.6 (absent of crystalline texture). The strong {100} sheet texture in two of the samples suggests that a recrystallization anneal was utilized, but the grains were not fully recrystallized. This is inferred since upon recrystallization annealing it has been reported that fcc metals that had undergone mechanical forming operations transform to the so-called cube orientation {100} <001> [23].

An interesting comparison among solid material forms may be made between the metallic sheet and metallic, inorganic, and organic powders. Several of these powders have shown evidence of texturing from XRD analysis during the course of our research. Unlike metallic sheet, the origin of their preferential alignment is largely from crystalline facets, flat geometric surfaces that reflect their underlying crystal structure, and result in organized settling of the particles. Figure 5.12 shows spectra for ammonium nitrate and sucrose powders. These plots are the mean of the diffraction spectra obtained at $\phi = 30°$ increments and acquired through a full 360° in-plane rotation; the error bars are the maximum standard deviation for the most prominent diffraction peaks. The reader is referred to Figure 5.2a for a description of ϕ angle in our diffractometer. While the spectral peak positions remain unchanged, the peak heights show variation as a function of ϕ angle due to crystalline texturing. There is no change in the crystal structure in the two materials which determines their spectra peak positions based on Bragg's law. But rather, the powders are configured in a preferentially oriented manner from their faceted nature giving rise to the peak height variability observed as a function of ϕ angle.

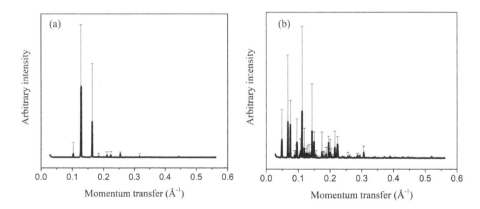

FIGURE 5.12
The X-ray diffraction spectra are the mean of spectra obtained from separate measurements at $\phi = 30°$ increments through a full 360° rotation for (a) ammonium nitrate and (b) sucrose. The error bars are the standard deviation for the most prominent diffraction peaks.

5.3 X-Ray Diffraction of Non-Crystalline Substances

In contrast to crystalline materials, amorphous solids and liquids possess only short-range order, lacking long-range atomic or molecular periodicity. The structural make-up of these materials can generally be described as disordered, isotropic, and homogeneous, which is reflected in their macroscopic physical properties [24]. For these materials, the causation of X-ray diffraction features requires a different physical model of the analyzing environment.

5.3.1 Non-Crystallinity and Short-Range Order

XRD phenomena leading to characteristic spectral features for amorphous substances is reliant on coherent scatter and interference as observed for crystalline materials but tied to short-range structural order in these materials. In this discussion, it is important to first acknowledge a distinction among materials types for amorphous solids. Based on thermodynamic criteria, two types of amorphous solids are definable, those that are non-crystallizable and those that are crystallizable [25]. The former defines ideal amorphous solids that are unable to crystallize under certain time-temperature-transformation conditions. The latter involves metastable materials that have large defect populations and a corresponding lack of crystallinity. It is noted that metastable phases may exist, a consequence of materials processing and chemical kinetics considerations. In fact, many times metastable materials are sought based on their desired properties. Nevertheless, diffraction phenomena from these various forms of amorphous solids and liquids are fundamentally the same. Were it not for short-range order in amorphous materials with particle (a generalized term representing atoms or molecules in amorphous materials) spacing on the order of X-ray wavelengths, structural characterization, and substance classification in X-ray diffraction imaging systems would not be possible.

The reproducible XRD features from amorphous solids and liquids give way to the means of determining spatial distributions of particles and, for purposes of threat detection, materials identification. Since short-range order dictates the structure of liquids and glasses, probabilistic atomic distribution functions are used to describe particle configuration [24]. The association of particles in liquids and amorphous solids is defined by the radial pair distribution function (RDF or, pair distribution function). RDF is used to determine bonding topology and the population of atoms in coordination spheres [26]—whereby a central particle is surrounded by an array of particles in the configuration of a sphere. Thus, pertinent data related to structure may be obtained from diffraction spectra. An example of coordination spheres is shown in Figure 5.13 for water, showing the distance r_1 from a reference molecule coincident with its first coordination sphere and r_2 for the second

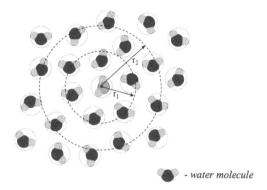

- *water molecule*

FIGURE 5.13
Illustration of water molecules and corresponding coordination spheres as indicated by the dashed lines. Distances from the reference molecule to each sphere are defined by a radial distance. The circles around each molecule signify their particle designation.

coordination sphere. This highlights the probabilistic nature of first-nearest neighbor distances, second-nearest neighbor distances, and so on. Nearest neighbor positions are defined by particle size (indicated in this figure by a circle around each water molecule) and repulsive intermolecular forces given by the properties of the material. The desire to maximize particle density while minimizing short-range repulsive forces leads to characteristic spectral features.

5.3.2 Spectral Features of Amorphous Solids and Liquids

It is instructive to consider diffraction features unique to amorphous solids and liquids. A differentiating aspect of these substances compared to crystalline materials is the form of these substances existing in either solid or liquid state. The structural construct of amorphous solids and liquids leads to their inherent homogeneity and isotropy, producing only a few broad, spatially uniform diffraction features. The observation of broad features is a consequence of a distribution in short-range spatial distances among particles in the sampling environment. Furthermore, the number of broad features is due to the ability of particles to configure into organized coordination spheres, as shown in Figure 5.13. The limited number of features and dramatic reduction in feature size with momentum transfer is due in part to a lesser probability of particles lying on larger coordination spheres.

The diffuse patterns for liquids arise from non-periodic configuration and unconstrained movement of solutions of atoms or molecules. The X-ray diffraction spectra features for water and ethanol in Figure 5.14 disclose chemical properties and spatial distance information. The spectrum for pure water shows three broad features tied to molecular configuration

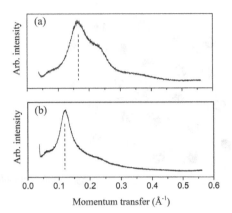

FIGURE 5.14
X-ray diffraction spectra for (a) pure water and (b) pure ethanol. The dashed line signifies the differences in the prominent peak positions for the two liquids.

and characteristics of water as reported in prior work [27,28]. The difference in the broad peak positions between water and ethanol is attributed to the larger ethanol molecule and correspondingly larger molecule-to-molecule spacing and larger radial distribution distance. Importantly, a clear distinction in spectral features for both can be made in this case for these aqueous and flammable liquids.

The diffraction patterns for amorphous solids look similar to amorphous liquids as expected given the absence of long-range order for both forms of matter. Figure 5.15 shows diffraction patterns for inorganic amorphous silica (SiO_2) and organic polymeric polymethylmethacrylate (PMMA). Similar to spectra observed for liquids, only a few, broad diffraction features are observed. The prominent diffraction feature for PMMA is at a lower momentum transfer value relative to the amorphous silica corresponding to the large size of the organic mer unit or molecular repeat unit, $(C_5O_2H_8)_n$. The PMMA spectrum also shows evidence of more structured coordination spheres through existence of three distinct broad diffraction features. Importantly, the spectrum for PMMA shows unique features relative to silica, pure water, and ethanol that allow this material to be readily identifiable. In contrast, the spectrum for amorphous silica is similar to that of ethanol. In this case, material density and effective atomic number from attenuation-based X-ray imaging could be used to distinguish these substances.

X-ray diffraction spectra of amorphous solids and liquids show similar attributes, characterized by few broad spectral peaks, as shown in Figure 5.16, for pure water and ethanol and mixtures of the two liquids. The peak intensity in this figure is related to the grayscale value (with darker features corresponding to larger peak intensities). The diffraction spectra can be used for substance classification due to unique particle configuration.

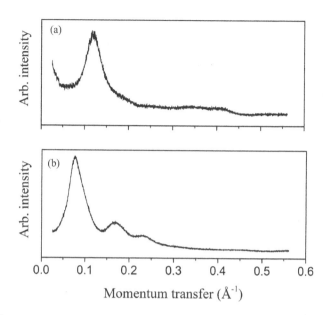

FIGURE 5.15
X-ray diffraction spectra for (a) inorganic amorphous silica and (b) organic polymeric polymethylmethacrylate.

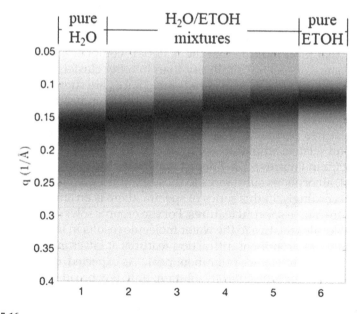

FIGURE 5.16
X-ray diffraction spectra, from left-to-right, for pure water (H_2O), and water/ethanol mixtures in increasing concentration to pure ethanol (ETOH, in increments of 20%). The brightness of the data indicates spectra intensity with black being larger and white being smaller intensities.

5.4 Substance Classification Based on Material-Science-Informed Cluster Analysis

There are clearly distinguishable diffraction features for crystalline and non-crystalline substances, as discussed thus far. Of interest are the shared spectral attributes among well-defined classes of materials, such as metals and alloys, polymers, ceramics, food stuffs, aqueous-based commercial goods, and so on, which may be used for substance classification purposes and highly relevant to the development of X-ray coherent scatter imaging systems. In this section, we exploit PSP relationships toward the objective of material-identification-free threat detection using cluster analysis, a statistical classification method [29,30] that we apply to stream of commerce goods using diffraction spectra.

5.4.1 X-Ray Diffraction Features for Classes of Materials

Cluster analysis is undertaken using an XRD library of 206 materials measured by the author and includes both threat (i.e., explosives and prohibited flammables, oxidizers, etc.) and non-threat (i.e., polymers, metals, foodstuffs, etc.) items. These materials can be distinguished into materials classes, possessing similarities in physical properties and elemental composition and material structure. Cluster analysis applied to X-ray diffraction spectra is an unsupervised learning method that compares intra-class vs inter-class spectra feature variability to group data according to relevant features. The data groupings that form indicate that the materials comprising a cluster have a higher measure of spectral similarity than in other clusters [31]. Diffraction spectra for several classes of materials/substances are shown in Figure 5.17 for flammable organic solvents, aqueous-based commercial goods, foodstuffs, polymers, metals and alloys, and solid oxidizer compounds. Key spectral features pertinent to cluster analysis include the width, number, location, and relative amplitudes of the diffraction peaks.

The spectra in Figures 5.17a–f depict materials classes that transition from disordered, amorphous substances to those exhibiting a high degree of crystallinity; accordingly, in this series of spectra, there is an apparent transition of smooth-to-course spectral features. For the organic solvents, the size of the organic molecules relative to the water molecule results in larger intramolecular spacing and prominent diffraction features at lower momentum transfer values relative to water's prominent peak. As expected, due to the lack of long-range order for molecules in solution, only few broad spectral features are observed. Importantly, these features are distinguishable from water features and have identifiable attributes that can be useful for substance classification. A strong water signature is observed as expected for aqueous-based commercial goods. Additional features are attributed to glycerol in lotion, shaving cream, and conditioner, and a broad peak at lower momentum

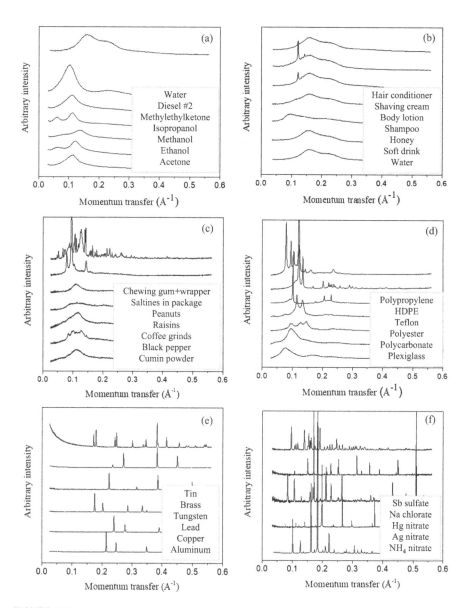

FIGURE 5.17
X-ray diffraction spectra for (a) organic solvents, (b) aqueous-based commercial goods, (c) food stuffs, (d) polymers, (e) metals and alloys, and (f) oxidizer compounds.

transfer values for honey from solubilized sugar. It is interesting to note the similarity among the food stuffs; even for vastly different types of foods, which indicate a similarity in the chemical make-up of these substances. The packaging for the chewing gum and saltine crackers contributes additional spectral features. Polymeric materials lack long-range crystallinity and, consequently, the diffraction spectra are easily distinguishable from those for metals and alloys. Because of only short-range molecular order, these materials do not configure into a Bravais lattice type and exist in varying proportions of quasi crystalline (from hydrocarbon chain organization) and amorphous structure. The diffraction features for metals and alloys are quite similar. Metals and alloys typical in commercial goods will exhibit strong crystallinity, characterized by low density, sparse diffraction features. The numerous peaks for the oxidizer compounds, originating at low momentum transfer values, are due to the lower symmetry crystal structures and the polyatomic ions residing on lattice sites producing large unit cells.

5.4.2 Unsupervised *K*-Means Cluster Analysis

Principal component (PC) analysis was used to perform a relative assessment among all materials in the library of XRD spectra including those in Figure 5.17. We found that normalizing each individual spectrum in the database such that the maximum peak value is set to unity produced the most intuitive results [32]. General observations from this analysis are that the first PC has relatively broad features while the second PC, third PC, and so on have increasingly more distinct, higher frequency features. This analysis was helpful in establishing the relationship between attributes of the principal components and those of the spectra features.

Inherent to *k*-means cluster analysis is its unsupervised machine-learning modality, aside from the choice of the number of clusters as a control parameter. The convenience of assigning *k* number of clusters is however interesting in our application as increasing numbers of clusters provide the opportunity for more meticulous materials structure classification. *k*-means clustering was implemented using MATLAB® software and was used to explore the relationship between the specified number of clusters and selected materials from the library of XRD spectra; all of the available spectral features were analyzed to determine the material groupings that emerged. Figure 5.18a shows a scatter plot projected onto the first and second PCs for $k = 2$ clusters, where the closed circles correspond to crystalline materials and the open circles to non-crystalline materials. There is a clear delineation between the data groupings indicating that our normalization procedure and application of only the first two PCs leads to obvious materials structure-based clusters. Figure 5.18b shows the average XRD spectra for the two clusters revealing that the cluster associated with the crystalline materials has more distinct, higher frequency peaks while the cluster associated with the non-crystalline materials has broader features. It is concluded that limiting the number of clusters to $k = 2$ distinguishes materials based on crystallinity, signifying the utility of XRD for structure determination.

We next considered $k = 7$ number of clusters to evaluate cluster grouping according to specific materials classes. In this scenario, classification by way of cluster formation would highlight the PSP relationship expected

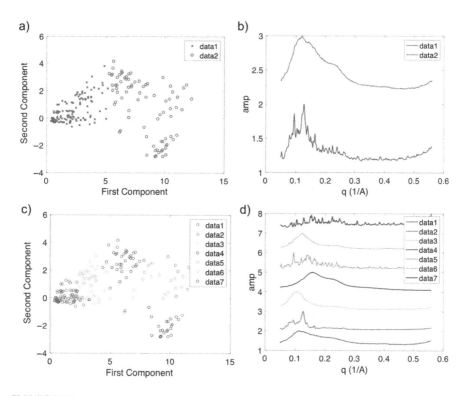

FIGURE 5.18

(a) Scatter plot for $k = 2$ clusters grouping according to data1: crystalline and data2: non-crystalline materials and (b) the average XRD spectra for the two clusters. (c) Scatter plot for $k = 7$ clusters grouping into data1: foodstuffs, data2: natural and synthetic polymers, data3: organics (amorphous polymers/liquid solvents), data4: water/weak acids, data5: crystalline explosives and flammables, data6: foodstuffs, and data7: sugar/vitamins/ammonium nitrate-based explosives and (d) the average XRD spectra for the seven clusters.

of materials classes. As shown in Figure 5.18c, distinct clusters based on classes of materials are represented in the scatter plot. These data clusters and associated materials classes are as follows: data1: foodstuffs, data2: natural and synthetic polymers, data3: organics (amorphous polymers/liquid solvents), data4: water/weak acids, data5: crystalline explosives and flammables, data6: foodstuffs, and data7: sugar/vitamins/AN-based explosives. Figure 5.18d shows the average XRD spectra for the seven clusters. It is noted that the average XRD spectra for data1: and data6: foodstuffs look similar, which may be due to both different compositions of the food as well as a mismatch between the intrinsic number of clusters and the choice of $k = 7$ used. The outcome of this cluster analysis indicates the great potential of X-ray diffraction imaging for substance classification without the need for an exhaustive library to perform direct comparison for the purpose of material classification. This point is critical since the variability and range of materials (and their associated form factors) leads to a nearly infinite range which would otherwise be impossible to capture fully in a single library.

5.5 Summary

X-ray diffraction imaging is well suited for aviation security applications leveraging a wealth of knowledge from materials characterization advances. Incident radiation interacts with objects to reveal details of atomic-level structure tied to materials properties which may be used for threat (or non-threat) assessment. Hence, the PSP relationship fundamental to materials science establishes the basis of this security technology. This chapter discussed X-ray diffraction features characteristic of crystalline and non-crystalline substances and the imaging modality for both of these material forms. Spectral anomalies were reported due to crystalline texture from processing and physical attributes of commercial goods. Statistical cluster analysis revealed the discriminating nature of X-ray diffraction for object classification based broadly on crystallinity for limited defined clusters and for more refined grouping of materials according to common materials properties for larger data cluster sets.

References

1. W.D. Callister, Jr., D.G. Rethwisch. *"Materials Science and Engineering: An Introduction"* (9th ed.), Hoboken, NJ: John Wiley and Sons, 2009, p. 5.
2. R. Boyle, *"The Sceptical Chymist or, Chymico-physical Doubts & Paradoxes"*, Printed by J. Cadwell for J. Cooke, (1661).
3. W.H. Miller, *"A Treatise on Crystallography"*, Cambridge: Printed at the Pitt Press: For J. & J.J. Deighton, (1839).
4. A. Bravais, "Memoire sur les systemes forms par des points distributes reguli-erement sur un l'espace", *Journal de l'Ecole Polytechnique* 19, 1 (1850).
5. N. Bohr, "On the Constitution of Atoms and Molecules", *Philos. Mag.* 26, 1 (1913).
6. E. Schrodinger, "Quatisierung als Eigenwertproblem", *Annalen der Physik* 384 (4), 273 (1926).
7. W. Röntgen, "Ueber eine neue Art von Strahlen. Vorläufige Mitteilung", in: Aus den Sitzungsberichten der Würzburger Physik.-medic. *Gesellschaft Würzburg*, 137 (1895).
8. R.W. Cahn, *"The Coming of Materials Science"*, Amsterdam: Pergamon, 2001 p. 67.
9. Harding, G. and Schreiber, B., "Coherent X-ray scatter imaging and its application in biomedical science and industry", *Rad. Phys. Chem.* 56, 229–245 (1999).

10. Luggar, R.D., Horrocks, J.A., Speller, R.D., and Lacey, R.J., "Low angle X-ray scatter for explosives detection: A geometry optimization", *Appl. Radiat. Isot.* 48(2), 215–224 (1997).

11. Strecker, H., Harding, G.L., Bomsdorf, H., Kanzenbach, J., Linde, R., and Martens, G., "Detection of explosives in airport baggage using coherent x-ray scatter", *Proc. SPIE 2092*, 399–410 (1994).

12. MacCabe, K., Krishnamurthy, K., Chawla, A., Marks, D., Samei, E., and Brady, D., "Pencil beam coded aperture X-ray scatter imaging," *Opt. Express* 20, 16310 (2012).

13. Greenberg, J. A., Krishnamurthy, K., and Brady, D., "Snapshot molecular imaging using coded energy-sensitive detection," *Opt. Express* 21, 25480 (2013).

14. J. A. Greenberg, M. Hassan, K. Krishnamurthy, and D. Brady, "Structured illumination for tomographic X-ray diffraction imaging," *Analyst* 139, 709 (2014).

15. W.H. Bragg and W.L. Bragg, "The reflexion of X-rays by crystals", *Proc. R. Soc. Lond. A.* 88 (605), 428 (1913).

16. A. Authier, *"Early Days of X-Ray Crystallography"*, Oxford: Oxford University Press, 2013, p. 319–376.

17. B.D. Cullity, *"Elements of X-Ray Diffraction"*, Reading, MA: Addison-Wesley Publishing Company, Inc., 1956, p. 314–316.

18. ICDD: The International Centre for Diffraction Data. Retrieved from www.icdd.com/.

19. B.D. Cullity, *"Elements of X-ray Diffraction"*, Reading, MA: Addison-Wesley Publishing Company, Inc., 1956, p. 301–304.

20. Suwas, S. and Ray, R.R., *"Crystallographic Texture of Materials"*, London: Springer-Verlag, , 2014, pp. 11–38.

21. Sowerby, R. and Johnson, W., "A review of texture and anisotropy in relation to metal forming", *Mater. Sci. Eng.* 20, 101–111 (1975).

22. Gottstein, G., "Evolution of recrystallization texture-classical approaches and recent advances", *Mater. Sci. Forum* 408–412, 1–24 (2002).

23. Lee, D. N. and Han, H. N., "Recrystallization textures of metals and alloys". In P. Wilson (Ed.), *Recent Developments in the Study of Recrystallization*, Rijeka: InTech, 2013.

24. H.E. Fischer, A.C. Barnes, P.S. Salmon, "Neutron and X-ray diffraction studies of liquids and glasses", *Rep. Prog. Phys.* 69, 233 (2006).

25. Z.H. Stachurski, "On structure and properties of amorphous materials", *Materials* 4, 1564 (2011).

26. J.H. Konnert, J. Karle, and P. D'Antonio, *"Radial Distribution Function Analysis"*, ASM Handbook Volume 10, Materials Characterization, R.E. Whan, Coordinator, ASM International, USA, 1986, 393–401.

27. J. Kosanetzky, B. Knoerr, G. Harding, U. Nietzel, "X-ray diffraction measurements of some plastic materials and body tissues", *Med. Phys.* 14(4), 526 (1987).

28. G. Harding, B. Schreiber, "Coherent X-ray scatter imaging and its applications in biomedical science and industry", *Rad. Phys. Chem.* 56, 229 (1999).

29. M.S. Aldenderfer and R.K. Blashfield, *"Cluster Analysis"*, New York: Sage Publications, 1985.

30. M.R. Anderberg, *"Cluster Analysis for Applications"*, New York: Academic Press, 1973.

31. Machine learning method for finding and visualizing natural groupings and patterns in data. MathWorks®: Cluster Analysis. Retrieved from www.mathworks.com/discovery/cluster-analysis.html.
32. S. Yuan, S. Wolter, and J.A. Greenberg, "Material-identification-free detection based on material-science-informed clustering", *Proc. SPIE 9847*, Anomaly Detection and Imaging with X-Rays (ADIX), 10187-20 (2017).

6

X-Ray Diffraction and Focal Construct Technology

Keith Rogers

Cranfield University

Paul Evans

Nottingham Trent University

CONTENTS

This chapter examines the background and practice of X-ray diffraction (XRD) and considers this phenomenon principally in the context of X-ray-based security screening. The focus will be upon the practical aspects of XRD as many texts already provide comprehensive descriptions of the relevant theoretical background and that of the closely associated area of crystallography. X-ray diffraction and its development from simple materials identification to dynamic imaging will be considered, followed by a similar view of aviation screening. Subsequently, a new approach to the harvesting of diffraction signatures (*Focal Construct Technology*) will be introduced and consequent potential applications summarised.

The text provides a starting point for those interested in developments that exploit X-ray diffraction, particularly in the security screening sector.

6.1 X-Ray Diffraction

Following the discovery of X-rays in 1895 by Roentgen [1], their exploitation produced two, almost irreconcilable, distinct disciplines: imaging (radiography) and X-ray diffraction. Von Laue, after demonstrating the wave nature of the rays, showed that the intriguing scatter patterns could be thought of as arising from a three-dimensional diffraction grating. When single crystals were illuminated by a pencil beam of X-rays, they produced discrete spots (intensity maxima) in regular patterns remote from the beam path on a photographic film. Von Laue [2] developed a series of relationships to interpret these patterns and they remain a cornerstone of crystallography today. Simply, and in one dimension, the condition for constructive interference from a row of coherently scattering centres separated by a vector distance, a, is that $a \cdot s = h$, where s is a scattering vector and 'h' an integer. For three-dimensional regular arrays of atoms, three such equations exist (the 'Laue equations') that fully describe the conditions for diffraction. The spatial distribution of the intensity maxima is related to a crystal's internal structure, i.e. the three-dimensional spatial arrangement of atoms. Structure determination was an early application for X-ray diffraction and remains the dominant activity for diffractionists and crystallographers [3]. A further interpretative breakthrough arose from Laurence Bragg's thesis that X-ray diffraction may be considered as simple specular reflection (at least in terms of direction) from sets of equi-spaced parallel lines intersecting with all the scatter centres as embodied within Bragg's law [4],

$$\lambda = 2d \sin \theta \qquad (6.1)$$

which establishes the simple relationship between interplanar spacing ('d', or 'd-spacing'), X-ray wavelength (λ), and direction of scatter relative to the incident beam, 2θ. For most diffractionists, perhaps with the exception of those studying large-scale structures using small angle scattering [5], only the first order of any atomic plane series is considered and therefore any integer order term is redundant. Bragg's law also indicates a key experiment condition that the X-ray wavelength employed for most crystallography is practically optimised when it is approximately that of the interplanar spacing, i.e. $\sim 10^{-10}$ m. In practice, Bragg's law illustrates elegantly the reciprocal relationship between Euclidian space (the separation of atomic planes) and diffraction space (θ); scattering occurs at relatively low angles for large-scale repeat distances.

Polycrystalline materials may be thought of as orientationally random collections of small (typically <100 μm) single crystals. Thus, when interrogated by a narrow pencil beam of X-rays, scatter follows the form of concentric hollow cones (Debye cones) shown in the forward direction in Figure 6.1. At normal incidence, these cones produce rings of relatively high intensity.

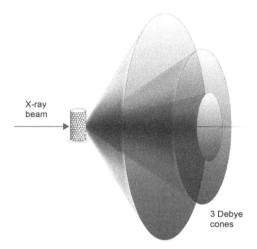

FIGURE 6.1
Idealized distribution of coherent X-ray scatter from a polycrystalline sample illuminated with monochromatic X-ray. Only three Debye cones are illustrated for clarity.

Under ideal conditions, the circumferential intensity is constant around each of the rings. Plotting the intensity against scatter angle produces an identical 1D graph regardless of the azimuth angle chosen. However, in practice, this ideal is met rarely; Figure 6.2 illustrates a conventional pencil beam system configured in transmission mode with a number of non-ideal scattering examples from samples with preferred orientation and large (relative to the interrogating X-ray beam) grain size. This type of data collection and its interpretation is the basis for all powder diffraction applications [6]. Plotting the relative intensity or photon counts over a chosen integration period against 2θ (i.e. the half-opening angle) of the different Debye cones produces a diffractogram.

To derive the atomic distributions within a crystal (and thus solve its structure), it is necessary to determine (a) the relationships between directions of scatter and (b) the amplitude of each maximum. Using the Fourier transform relationship, expressed in Eq. (6.2), the electron density distribution, $\rho(x, y, z)$, may be calculated from the scatter amplitudes, F_{hkl} if the phase, ϕ_{hkl}, of each reflection (intensity maxima) is known [7].

$$\rho(xyz) = \frac{1}{V} \sum_{h=-\infty}^{h=\infty} \sum_{k=-\infty}^{k=\infty} \sum_{l=-\infty}^{l=\infty} |F_{hkl}| e^{i\phi_{hkl}} e^{-2\pi i(hx+ky+lz)} \tag{6.2}$$

where h, k, l are Miller indices defining a set of planes, V is the unit cell volume and x, y, z are unit cell fractional coordinates.

However, as the phase cannot be determined directly from measurements of intensity, a direct calculation of electron density distribution is not

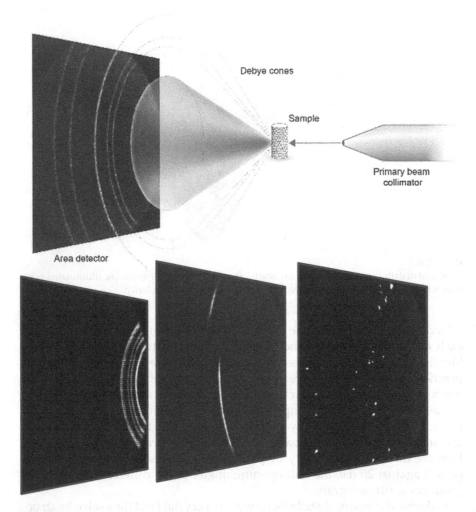

FIGURE 6.2
Upper diagram illustrates a typical transmission diffraction experiment where a monochromatic primary beam illuminates a polycrystalline sample, and a section of the coherent scatter is captured upon a planer area detector. In this case, the pattern recorded illustrates that from an 'ideal' sample. The lower diagrams illustrate corresponding scattering from materials with significantly greater lattice parameters (left), pronounced preferred orientation (middle), and relatively large grain size (right).

possible from experiment. Many methods have been developed to solve this 'phase problem' and the reader is referred to excellent texts for further discussion [8]. It becomes apparent when studying the distribution of intensity maxima that atomic planes may appear to possess scattering amplitudes equal to zero; they are absent from the diffractogram. This result arises due to the specific inherent symmetry of regular atomic arrangements; determination of the symmetry components is a critical step in determining structures. The structure determination process is the dominant activity for diffractionists and crystallographers and has led to several celebrated discoveries including the elucidation of the DNA structure [9] and that of haemoglobin [10] where structure–function relationships have been shown to be critical to our understanding of biochemistry. Structure solution of proteins is now almost routine with perhaps the most challenging component being the crystallisation stage. Phase retrieval methods can be applied to data from single crystals and also polycrystals although powder diffraction is described by a one-dimensional diffraction space (i.e. planes of the same family with identical *d*-spacings are generally indistinguishable from one another) and thus single crystal diffraction is the preferred choice especially for large molecular structure elucidation. The historical limitation that many materials do not easily form large single crystals is rapidly being superseded with the combination of bright X-ray sources (especially synchrotrons) and environmental control enabling structure solution from single, micron-sized crystals [11].

In contrast to structure determination, materials identification from diffraction signatures is most often undertaken using polycrystalline materials with reference to a database of 'known' (empirical or calculated) diffraction patterns [12]. The term 'phase' is used in this context to describe the unique structures of a material species. For example, calcium carbonate may form as calcite or aragonite structural phases. Interestingly, a significant biological mystery is how cells engineer the formation of one phase rather than the other [13]. Although the stoichiometry of $CaCO_3$ is identical for calcite and aragonite, their atomic arrangements are quite different and thus they possess very distinct diffraction patterns as illustrated in Figure 6.3. Thus, one of the strengths of diffraction for materials discrimination is its ability to discriminate between such 'polymorphs'.

In principle, the process of materials identification involves recording the directions of scatter and relative intensities for each Bragg maxima and then comparing them with those of previously collected data. This is a 'fingerprint' approach to identification that originated through systematic procedures developed, for example, by Hanawalt [14]. The first database, or 'powder diffraction file' (PDF), was produced in 1941, and this has developed into a significant (>384,000 entries) commercial database, 'PDF4+', available from the International Centre for Diffraction Data (ICDD) [15].

For several years, powder diffraction languished as a technique in the shadow of single crystal methods principally due to the problems of extracting accurate integrated intensities from overlapping sets of intensity maxima. However, something of a revolution occurred when Hugo Rietveld showed how each individual intensity point within a dataset could be exploited [16]. His method involves a non-linear least squares fit of the experimental data points to those calculated from an initial or prototype crystallographic structure. The 'Rietveld refinement method' [17] has now become a ubiquitous tool and, despite producing parameter standard deviations that possibly

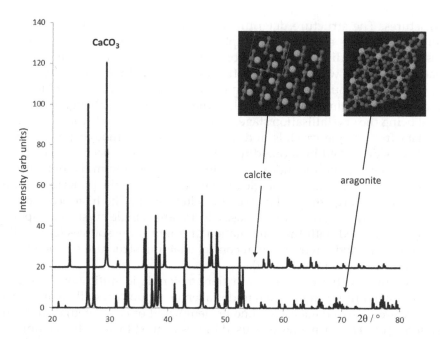

FIGURE 6.3
Two polymorphs of calcium carbonate illustrating how identical chemical stoichiometry can result in significantly different diffractograms.

underestimate probable errors, it continues to be a very successful approach to structural refinement.

Contemporary analysis methods applied to data from powder diffraction experiments enable derivation of features well beyond simply material phase. For example, it is possible to extract microstructural, quantitative information such as micro-strain, macro-strain, coherent domain size (crystallite size), and preferred orientation [6]. Indeed, any physical characteristic that can be mathematically modelled can be included within a Rietveld refinement and quantified. This has led, for example, to standardless quantification of phases within a mixture, i.e. without reference to external calibration curves [18]. Consequently, powder X-ray diffraction has become a powerful tool in the materials analysis armoury.

In contrast to crystalline materials, amorphous materials (including liquids) do not possess long-range structural order and therefore do not produce scattering distributions with well-defined (sharp) diffraction maxima. However, a degree of short range, local order does persist (e.g. from inter-atomic and intra-molecular bonds) and thus corresponding scattering distributions are characterised by broad, diffuse halos [19]. Analytically, these are considered through radial or pair distribution functions that describe the average atomic or electron density as a function of the radial distance from any reference atom; interatomic vector directions are meaningless within a structurally disordered system, only their magnitudes have relevance. Thus, the scattering intensity distribution $I(q)$ from a non-crystalline array of atoms may be described by:

$$I(q) = \sum_m \sum_n f_m * f_n \sin qr_{nm}/qr_{nm} \qquad (6.3)$$

where $q = 4\pi \sin \theta/\lambda$, r_{nm} is the magnitude of the vector separating atoms m & n, and f_m & f_n are the respective atomic scattering factors for each atom.

For example, water produces a dominant broad peak with a maximum at ~0.324 nm [20] that arises from the intermolecular interference of the nearest neighbour oxygen–oxygen (O–O) atoms. Examples of X-ray scattering from liquids is provided in Figure 6.4.

In parallel to analytical innovations, developments in hardware have enabled elucidation of increasingly complex molecular structures and have reduced data collection times. For example, digital area detectors, including those with pixelated energy-resolving capability [21], have replaced photographic film, and synchrotrons have become the X-ray source of choice for many experimentalists especially those wishing to achieve high-spatial resolutions or observe stimulated structural changes dynamically [22]. Further

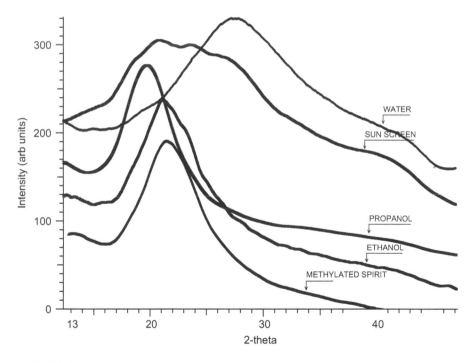

FIGURE 6.4
X-ray coherent scatter distributions (monochromatic) from a range of materials in liquid form.

the use of engineered optical elements (e.g. graded layers, polycapillaries, etc.) has also enhanced the diffraction experiment [23].

For routine, laboratory-based powder diffraction in reflection mode, the sample is illuminated with a collimated, monchromatised X-ray beam, and a moving detector rotates as illustrated in Figure 6.5. As dispersion occurs spatially, this is known as angular dispersive X-ray diffraction (ADXRD). Conventionally, the Bragg–Brentano parafocusing geometry is employed and the sample is rotated around an axis parallel to the detector rotation axis to maintain a fixed $\theta/2\theta$ relationship [6]. This reflection geometry results in only those planes parallel to the sample surface being measured. The sample may also be rotated around an axis normal to sample plane to improve counting statistics.

The choice of interrogating wavelength is an important feature of any diffraction experiment. Typically for thermionic X-ray sources, a compromise is found between the anode melting points (high-power density & inefficient energy conversion), cost, and emission line energy. The CuKα line (0.154 nm) is the most frequently used and produced by the monochromatisation of radiation from a Cu anode. However, the energy of this radiation results in very little penetration into materials. For example, 99% of the X-ray photons contributing to the diffraction pattern of rhombohedral Al_2O_3 arise within the uppermost ~52 μm of the sample surface. This lack of penetration limits the use of diffraction for applications involving extended absorption paths encountered within practical objects, in particular, aviation security (see later).

An alternative, analogous method of data collection is to employ energy dispersive diffraction (EDXRD) where the sample is illuminated with a

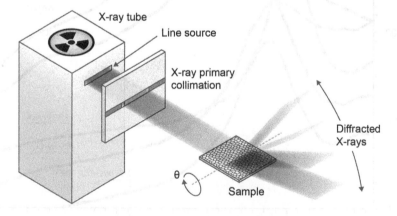

FIGURE 6.5
A conventional, reflection mode, powder diffraction experimental arrangement. A line source is collimated to direct a 'ribbon' of X-rays onto a flat plate sample, and a detector rotates around the sample to receive the scattered photons. Monochromatisation occurs using a filter or highly reflecting crystal placed within the incident or diffracted radiation (not illustrated).

polychromatic beam and an energy-resolving detector placed to receive dif-fracted radiation at a fixed position, θ [24]. Considering Bragg's law, the inter-planar spacing can then be calculated from the wavelength of the intensity maxima. In contrast to ADXRD where data is most frequently presented in terms of 2θ, EDXRD employs 'momentum transfer', i.e. $E \sin \theta / hc$. In prac-tice, data can be measured at significantly greater speeds using EDXRD. However, it is not often the method of choice as the d-spacing discrimination is compromised by the energy resolution and finite acceptance angle of the detector. Thus, when applied within a typical materials identification pro-tocol, the EDXRD approach produces significantly less confidence than that of the corresponding ADXRD data. This limitation impacts detrimentally upon the discrimination power of such systems particularly for materials with larger unit cells.

As a testament to the success of X-ray diffraction, the technique has remained (albeit with some refinement) essentially unmodified for over 100 years. The relatively high resolution (in the sense of d-spacing discrimi-nation) and other advances now enable structure determinations from poly-crystalline samples [25]—an achievement previously confined to the domain of single crystal crystallography. Furthermore, the growth in X-ray powder diffraction can be evidenced by diffractometer and database sales that have been annually double digit for the past decade. In summary, X-ray diffrac-tion is a powerful probe enabling unprecedented materials discrimination, identification, and characterisation. However, for potential applications that require high-speed data acquisition and perhaps the interrogation of materi-als requiring substantive X-ray penetration (most ADXRD experiments are performed in reflection mode), XRD in its current form is limited. High-speed/large area detectors and high-power sources may be prohibitively high cost for some industrial applications [26]. By employing a source of parallel X-rays, the illuminated footprint on the sample can be extended from a small area to a larger rectangular area to improve the signal-to-noise ratio, speed, and particle statistics. The usefulness of this approach is limited since the detector active area must be expanded with the X-ray footprint, resulting in a concomitant increase in the measured incoherent scatter. Furthermore, this approach causes peak asymmetry, particularly at low-scattering angles.

6.2 Aviation Screening

Over the last four decades, there has been a growing worldwide interest in aviation security technologies. This interest has been driven and directed by politics, technology developments, and terrorist threats. Many innovations arise and are implemented as a direct reaction to specific events, several of which are illustrated in Table 6.1. Between 1985 and 1997, around 1,100 people

TABLE 6.1

Aviation Terrorist Events

Year	Event	Reaction
1930	PanAm hijack	
1955	UA terrorist explosion	
1960	NA terrorist explosion	
1961	NA hijack	Armed guards on commercial flights
1969	TWA hijack	FAA metal detectors
1970	PanAm & TWA terrorist explosion	Sky marshals
1971	Northwest Orient hijack	Passenger screening
1972	TWA bomb threat	Explosives Detection Canine Team Program
1973	PanAm & Lufthansa hijack	
1974	Explosion at LAX	Adoption of metal-detectors and X-ray screening for passengers and carry-on bags
1985	TWA hijack	Federal air marshals
1988	PanAm terrorist explosion	Matched search to passengers
1997		Sig uplift in security funding
1998		Pre-screening systems employed
2001	AA & UA terrorist atrocity	TSA formed
2001	Shoe bomber	Shoe inspections, items banned
2002	Gunman at LAX	TSA screens all checked bags
2004		TSA Registered Traveller program
2006	Plot involving liquid explosives	All liquids divested from bags
2009	Northwestern passenger with explosive in underwear	Full-body scanners

lost their lives due to aviation terrorist bombings. The tragic events of 9/11, when almost 3,500 died, further increased global awareness and the demand for development of improved screening techniques.

Prior to the Lockerbie event in 1988, airport security concentrated on the detection of high atomic number objects such as weapons including guns and ammunition through the employment of metal detectors and single view X-ray screening. Subsequently, events involving passengers carrying explosive materials directed technology development towards detection of lower atomic number substances in highly shape variant formats. A significant issue has been identifying and adapting detection systems to deliver the necessary performance fidelity within the operational constraints demanded by the industry. A viable technology has to offer a cost-effective and preferably small footprint design, which is capable of being integrated into a high passenger throughput environment. Furthermore, the London transatlantic bomb plot of 2006 and the 'printer cartridge bomb plot' of 2010 are evidence for the requirement of systems to be capable of identifying liquids, aerosols, and gel explosives (LAGs), concealed home-made explosives (HMEs), respectively. In particular, the sophistication and inherent variability within the makeup of HMEs and their concealment is also increasing. The growing uncertainty associated with this threat vector is driving the demand for technologies with high throughput, sensitivity, and specificity to meet the required low false negatives and false alarm rates.

Both bulk and trace methods of explosives detection are components of aviation security screening systems. Explosive trace detection (ETD) techniques include mass spectrometry [27], chemiluminescence [28], ion mobility spectrometry [29], immunoassay [30], and bio-sensor technologies [31]. In general, these detect minute concentrations of an illicit substance (<1 mg) present on the exterior surface of luggage or vapours emitting from the substance. The primary detection ETD methods are inherently slow processes and produce an unacceptably low throughput for busy airports. In contrast, bulk detection systems, usually referred to as explosive detection systems (EDS), identify weapons and/or volumes of illicit substances, such as explosives and drugs, whilst screening 100% of checked baggage. These methods are frequently X-ray based and capture images of the inside of luggage. Other non-X-ray based screening systems include neutron techniques [32], nuclear quadrupole resonance [33], and terahertz time domain spectroscopy [34].

X-radiographic transmission screening is ubiquitous throughout the aviation security sector. The simple, single view systems have been enhanced by the addition of multiple views (providing some depth decoupling) and dual energy X-ray detectors enabling some material discrimination. However, there remain challenges for these technologies. For example, materials in sheet form are difficult to detect and the discrimination of materials based upon density/absorption characteristics is too inaccurate to prevent high levels of false positives.

The detection of sheet materials can be achieved through computer tomography (CT) where volumetric representations of objects are mathematically reconstructed from multiple views. Translation of this technology from the medical arena to the aviation security sector has been possible due to improvements in processor and data acquisition speeds. CT systems are commonly employed as a level-2 intervention, i.e. post-conventional X-ray scanning or in cargo screening. CT may also be coupled with dual energy detectors to produce greater, material discrimination, and thus reduced false alarm rates [35]. The cutting-edge RTT™ systems of Rapiscan are fast (scan speed of 0.5 ms⁻¹) and feature a novel stationary gantry. However, disadvantages of CT systems include their relatively high cost (capital and maintenance) and large footprint. Furthermore, regardless of superior imaging performance, sufficient material discrimination fidelity remains a significant hurdle to achieving low false alarm rates.

X-ray diffraction may be considered as providing orthogonal material information to that of absorption-based processes. As discussed previously, diffraction signatures are unique within materials space and could in principle be used for high-specificity detection and identification. This feature has promoted several attempts to exploit diffraction in security screening although there is only one notable commercial system incorporating a diffraction probe currently available. The 'XRD 3500™', commercialised by Morpho (Safran) is employed in a 'system of systems' approach as a secondary screening technique, upon identification of suspicious materials within an inspection volume by CT. It is a relatively high cost system (GPB ~ 0.5 M) requiring substantive maintenance. It employs high-power X-ray sources to increase the amount of diffracted flux available for analysis. Any discussion of XRD applied within aviation screening would not be complete without consideration of the work of Geoffrey Harding who has consistently promoted diffraction-based imaging systems through numerous elegant patents and publications. For example, the multi-generational X-ray diffraction imaging (XDI) technique was introduced in 2005 [36] as a concept system for security screening, combining the ability of X-rays to form an image and to analyse the material under inspection. This was superseded by the 'multiple inverse fan beam' (MIFB) topology [37].

Despite these innovations, practical solutions that enable the deployment of XRD within mass transit screening systems have remained elusive. The principal barrier has been the relatively small amount of diffracted flux collected during operationally relevant exposure times leading to low signal-to-noise ratio. The relatively small amount of signal collected is usually due to the compound effect of high aspect ratio collimation (photons collected over a small solid angle) and the low exposure times required for operationally relevant speeds. Consequently, many approaches require higher powered X-ray sources to increase the probability of collecting diffracted photons in order to make a material call.

6.3 Focal Construct Technology (FCT): Background

Traditional diffractometer methods measure the intensity and angular distribution of Debye cones from a sample using a point detector that cuts through each cone footprint or ring. In practice, point detectors have a finite detection area, and the intensity over a small, approximately annular sector of each ring is measured. The remaining fractions of the rings are usually not examined even though important data can be collected by measuring a whole ring. To measure complete Debye rings requires a spatially resolving area detector, which can be bulky and prohibitively high in cost. Furthermore, there is a trade-off between d-spacing resolution and sample to detector separation. For example, to capture rings over a wide range of Bragg angles requires either a very large area detector or a smaller detector positioned closer to the sample, i.e. to conserve the solid angle under consideration. In such a scenario, with fixed detector parameters, the larger detector will provide improved angular resolution and better estimation of d-spacing values. In conclusion, the layout and relatively large physical size of standard instruments such as powder diffractometers are dictated by the divergence of the diffracted flux from the samples necessitating off axis measurement.

Focal construct geometries employ a novel tubular interrogating or primary X-ray beam to affect a convergent and therefore inherently compact, diffracted ray geometry [38]. Whilst increasing the interrogated or gauge volume in comparison with a pencil beam, this technique produces diffracted rays that converge to a single series of collinear points [39], i.e. on axis measurement. To illustrate the basic focal construct concept in angular dispersive mode, consider a simple diffraction experiment where a pencil X-ray beam (diameter, W_T) strikes an ideal powder sample and produces (for simplicity in transmission) a 'single' Debye cone. If multiple incident beams are incident normally upon the sample such that the loci of their intersections with the sample lie equidistantly on a circle (radius, R_s) and the sample to detector distance, $D_z = R_s \cot 2\theta$, then the resultant Debye cones will possess a single convergence point along the principal axis of the interrogating tubular beam. This concept is illustrated in Figure 6.6 where only two incident beams and two corresponding Debye cones are shown for clarity. Two spatially separated 'focus points' are formed along the principal axis, z. Similar focus points are also recorded for a conical shell incident beam (although the equation above becomes modified). Thus, forming a continuous, annular interrogating X-ray footprint on the sample results in a high-intensity point along the principal axis (for each scattering plane) and an increased intensity

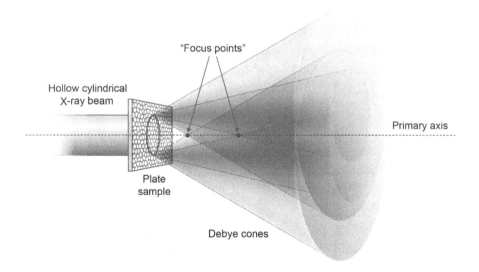

FIGURE 6.6
The principle of focal construct technology illustrated for just two points on the continuum of the annular incident beam (at the sample) and two resultant Debye cones. The 'focus points' occur along the primary axis.

circle or caustic at the outer Debye ring terminus with a radius, $\sim 2R_s$. Such loci occur due to the superposition of scattering from a continuum of points from a single family (constant two thetas) of Debye cones.

Systematically translating an area detector along the principal axis records a scattering distribution as illustrated in Figure 6.7. Interestingly, as the detector approaches a 'focal' position, it is clear that the extraneous scatter is significantly reduced. Summing the detector pixel intensity in the central region as a function of detector positon results in a 1D diffractogram with familiar high-intensity peaks [38]. However, the circular caustics may be equivalently formed from converging or diverging Debye ring rays and thus the abscissa cannot be straightforwardly transformed into scatter angle or d-spacing.

There are several advantages of this technique when compared to the traditional pencil beam experiments. The summation of intensity from any Debye family at the detector will provide order(s) of magnitude increase in intensity dependent upon design specification, i.e. intensity will be greater by a factor of $\sim 8R_s/W_T$ in comparison to an equivalent pencil beam measurement with a diameter, W_T. Furthermore, the detrimental effects of preferred orientation (e.g. wire texture) and large grain size on the diffraction data

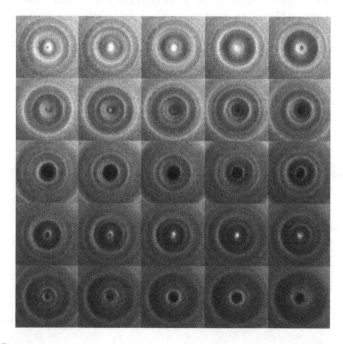

FIGURE 6.7
A sequence of X-ray scatter collected as a planer detector is systematically translated along the primary axis. The focal points are apparent from a central bright intensity point corresponding to scattering from particular sets of crystallographic planes. These Bragg maxima occur when the detector-sample distance is equal to $R \tan 2\theta$ (assuming parallel incident radiation).

may be reduced significantly without the need for any sample preparation [40]. FCT patterns corresponding to materials with preferred orientation and large grain size are illustrated in Figure 6.8 for detector positions at a focal point and also on either side of this maxima position. There are several modes of operation including the linear measurement of intensity along the principal axis and static energy dispersive detectors and, in principle, system elements may be engineered using standard, low-cost techniques. Typical *d*-spacing ranges can be measured within a short distance but with resolution comparable to that of conventional angular dispersive systems, i.e. there is a potential for smaller/portable implementation. In contrast to conventional approaches, here the greatest discriminating power occurs for high *d*-spacings. Any background signal (e.g. fluorescence) relative to the diffraction signal will be significantly less than that of a conventional system (possessing an equivalent illuminated area) simply due to the smaller detector area required to capture the coherent scatter.

There are a number of approaches to formalise the scattering process from this geometry. The scattering distribution at the detector may be considered

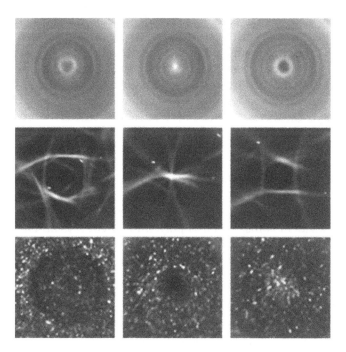

FIGURE 6.8
Focal construct coherent scatter distributions for materials measured with the sample-detector distance set to satisfy a focus condition (central column) and a distances slightly less (left column) and more than (right column) the focus condition. The upper row illustrates scattering from an 'ideal' polycrystalline sample (Al_2O_3), the middle row illustrates scattering from a sample with significant preferred orientation (Al) and the lower row illustrates scattering from a sample with large grain size ($KClO_3$). The focal condition is apparent in all three rows.

as a convolution of the Debye cones and the annular incident beam. Adopting Gaussian profiles for the Debye cone, $g(r)$, and incident beam, $f(r)$, cross sections (and assuming no beam divergence and an infinitely thin sample), then, for a single Debye cone the scattering distribution normal to the primary axis is represented by [40],

$$f(r)*g(r) = \frac{1}{\sqrt{2\pi\left(\sigma_f{}^2+\sigma_g{}^2\right)}}e^{\frac{-\left(r-(R-D\tan 2\theta)\right)^2}{2\left(\sigma_f{}^2+\sigma_g{}^2\right)}} \tag{6.4}$$

where σ_f & σ_g are the Gaussian width parameters for the incident beam and Debye cone, respectively, D is the perpendicular sample to detector distance, and R is the radius of the incident beam at the sample.

This is a continuous function of both r and D, and by plotting $f*g$ against D at $r=0$, the intensity profile along the principal axis can be calculated. By differentiating this with respect to 'D', the focal point positions along the principal axis can be derived and shown to be entirely coincident with the geometric derivation of focal points provided previously. This approach can also be applied to predict the scattering distributions from non-ideal samples such as those possessing preferred orientation.

Fortuitously, this equation has a form similar to that of a desirable autocorrelation function as provided by coded apertures such as MURA's [41]. In this respect, the beam topological modulation produced by the primary optic may be regarded as a coded aperture. This then has the advantage that the typical FCT ring caustics formed from both diverging and converging scatter cones (making them difficult to interpret as a diffractogram) may be accurately recovered through convolution.

Although angular dispersive FCT enables a study of 3D scattering distributions, the corresponding energy-dispersive approach has shown great promise in providing data acquisition at relatively high speeds [42]. The method exploits an incident conical shell of polychromatic X-rays and a single point or pixelated energy-resolving detector. It has also been shown capable of providing 'single shot' diffraction tomography (transmission) for materials identification through barriers (see later).

To date, there has been little exploration of FCT when applied in reflection mode analogous to that of conventional X-ray diffraction experiments. However, the 'focusing' advantage of the annular beam geometry can also be realised if the incident beam topology is such that it produces a specific footprint shape formed upon an inclined sample surface. In energy-dispersive mode, such a topology can produce a 'focal' point on the same side of the sample as the X-ray source and with all diffracted ray paths possessing the same scatter angle. This may be thought of as a 2D Bragg–Brentano geometry. To achieve such a focus, the X-ray source and detector are placed at the same height, h, from a planar sample, and separated by a distance D, and the beam topology would be such that an ellipsoidal footprint is illuminated on the sample surface. A powerful feature of this approach, unlike the conventional Bragg–Brentano method, is that that extended planar samples do not degrade the focal spot and there is no requirement for a curved sample.

All the FCT modes above require the formation of a hollow cone of X-rays and several practical options are available for this from a simple annular collimator (low-cost, easy construction but potentially poor uniformity and low intensity), to multilayer diffraction optic elements producing converging X-ray annuli [43]. The potential advantage of this particular approach is that it forms simultaneously converging and diverging X-ray beams.

In summary, FCT is a unique approach to the acquisition of X-ray diffraction data from polycrystalline materials and has inherent speed advantages that are particularly attractive for a number of industrial applications.

6.4 Focal Construct Technology: Applications

In principle, FCT could impact a wide range of sectors that currently exploit X-ray diffraction. However, we shall confine the discussion below to those areas where conventional powder diffraction methods are not ideal and where some progress using FCT has already been made. Currently, there are no commercial systems exploiting FCT, thus we shall consider mostly research data that will form the foundation of future developments.

6.4.1 Security: Aviation Screening

The principal advantage of FCT applied to aviation security concerns the increased relative intensity (compared to a pencil beam), which translates into a signal acquisition speed that meets the demands of current airport screening. In a high-energy (~150 kV) dispersive mode, FCT also satisfies the requirements for penetration through at least carry-on luggage. A further advantage of FCT is that both the absorption and diffraction caustics may be used in a tomosynthesis approach to directly image objects [44,45]. This is analogous to conventional CT scanning where the object appears to have been rotated about an axis normal to the detection surface in an oblique beam formed from a composite of annular projections. This is an interesting and counter-intuitive result as the conical shell incident beam is produced using a point source, and only linear motion is employed in the image collection process.

The initial application of FCT within a screening environment couples a pre-screening element with a 'point-and-shoot' FCT probe. This allows operators to resolve potential material ambiguities that would otherwise be considered threats. In particular, the diffraction probe is able to identify HME materials with a high degree of accuracy. Figure 6.9 schematically illustrates

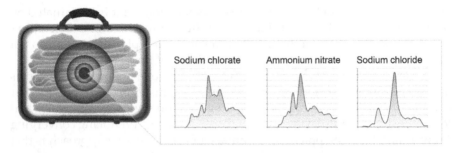

FIGURE 6.9
A schematic illustrating early results from FCT applied to identify material samples placed
within the cluttered environment of a suitcase. Confounding scatter from the suitcase, benign
contents, and other crystalline components were accommodated within the methodology and
each material identified with a high degree of specificity from its diffraction pattern.

some initial results from this probe. A relatively small mass of 'target' material was placed randomly in a suitcase within a cluttered environment and
examined with the FCT system. Diffraction signatures from three crystal-
lographically different materials produced using a 10 mAs exposure are also
illustrated in Figure 6.9. These materials were easily and uniquely identified
from the d-spacings (energy) of the Bragg maxima to demonstrate the high
accuracy materials discrimination.

6.4.2 Security: Envelopes and Packages

Conventional mailroom X-ray inspection equipment may assist in the detection of suspicious items from their shape using the differential absorption
processes of radiography. This inspection task can be slow and cumbersome
and requires training to interpret correctly 'threat' shapes. Of course, this
approach is not appropriate for threats such as explosives and narcotics where
shape is irrelevant and X-ray absorption features are poorly discriminating
[46]. Government agencies, including the US Postal Inspection Services and
the UK Home Office, are acutely aware of this problem but are forced to
rely on unsophisticated 'shape-based' technologies. Perhaps the three principal advantages of FCT are its interrogating speed, its discriminating ability, and also its potential for automation. Furthermore, as discussed above,
FCT lends itself to compact, reduced cost, and high-speed X-ray diffraction
instrumentation. Although FCT-based commercial systems are not currently
available, initial indications of the materials identification capability of FCT
for illicit materials within envelopes are positive.

6.4.3 Medical Diagnostics

The human body is a highly complex and dynamic assembly of biological,
chemical, and 'engineering' components. Medical diagnostics attempt to

identify, at the earliest stages, when any or all of these components demonstrate atypical behaviour. Unfortunately, due to intrinsic biological variability, such behavioural deviations are often difficult to identify and this often results in diagnoses at late stages of disease progression. Furthermore, diagnostics are often based upon observation of a symptom rather than an original cause. Although the advent of gene mapping has produced a revolution in medical diagnostics [47], it is often unable to make highly accurate predictions and, naturally, is almost redundant for diseases with environmental causes.

X-ray diffraction is not often associated with medical diagnostics. However, its ability to probe structural scales from hundreds of nanometres to Angstroms means that it can characterise both large molecular tertiary structures (such as those formed from fibrous proteins e.g. collagen) and the atomic arrangements in inorganic components, such as apatites in bone. Using ionising radiations is an acceptable practice in medicine when the diagnostic benefits outweigh the exposure risks. FCT's inherent ability to accelerate the data acquisition process has the potential to maintain a low patient radiation dose and further, its innate averaging of interrogation volumes militates against local, sampling inhomogeneity. The subsequent discussion concerns two areas where research is currently being undertaken.

Breast cancer is the most common cancer in women worldwide, and in 2012, almost 1.7 million new cases were diagnosed [48]. The ability of X-ray diffraction to probe wide architectural scales has been shown to enable the early-stage diagnosis of breast cancer [49]. This arises from the discovery that type I/III collagen suprastructure associated with breast tissues becomes significantly compromised as cancer invades surrounding tissues. Of particular interest is that this effect can be observed some distance from the tumour site thus suggesting the possibility of defining tumour margins using this structural characteristic. More recently, X-ray diffraction has also been shown to have diagnostic abilities capable of distinguishing benign and invasive breast cancers [50]. Although patient numbers at this time are relatively small, researchers identified systematic differences in the phase compositions and precise chemistry of breast calcifications (perhaps the most radiologically important early diagnostic markers of breast cancer) by XRD and optical spectroscopies [51]. Thus, characteristics of both soft tissue and calcification revealed through diffraction signatures may be thought of as constituting a new type of biomarker for breast cancer. Access to such biomarkers in vivo is a significant challenge but given the characteristics of FCT this technology has the potential to play a significant role within this field.

Osteoporosis affects 200 million women worldwide – approximately 75 million people in Europe, USA, and Japan [52]. It is estimated that an osteoporotic fracture occurs every 3 seconds. This is a condition that produces a major health burden and a mortality risk of between two and ten in the 12 months following a hip fracture. These issues are compounded by the asymptomatic nature of osteoporosis, which often remains undiagnosed until after a fracture occurs. As an engineering problem, one might compare bone mechanical functionality to that of an engineering structure such as a bridge. There are then three principal factors influencing mechanical performance: the architecture, the mass of construction material, and the composition ('quality') of the construction material. The first two of these are accessible through conventional imaging and bone mineral density (BMD) assessments determined by dual energy X-ray absorptiometry (DEXA) [53]. However, it has been consistently demonstrated that BMD alone is a poor predictor of fracture. Data from the Study of Osteoporotic Fractures (NIH: results on-line) showed that 54% of new hip fractures occurred in women who did not have osteoporosis as determined by their BMD, and data from the National Osteoporosis Risk Assessment showed that 82% of post-menopausal women with fractures had bone of 'normal' BMD.

The physicochemical characteristics of bone mineral (similar to calcium hydroxyapatite) are regularly examined by X-ray diffraction [54]. Bone is a composite material wherein the mineral component contributes to bone stiffness and strength, whereas collagen affects toughness. Features such as the degree (and to some extent the type) of lattice substitution, the coherent domain size, microstrain, and crystallite dimensions can be determined from diffractograms of native bone [55]. These are critical characteristics that contribute to the bone 'quality' and therefore impact on its mechanical performance. For example, mineral crystals possessing high levels of carbonate (exchanged for phosphate) are known to be preferentially resorbed under physiological conditions, and therefore if crystallites are precipitated with relatively high amounts of lattice-bound CO_3^{2-}, then bone mass is likely reduced [56]. A similar argument can be proposed for crystallites formed with a smaller domain size. Recently, studies have indicated that there are indeed significant differences in several physiochemical features of bone from patients with and without fragility hip fractures [57]. X-ray diffractograms were processed through parameterisation and also data mining methods (e.g. principal component analysis). The data mining approach indicated high accuracy of the data to predict patients within the fracture group [58]. Thus, bone diffraction signatures show great promise as biomarkers for osteoporosis. As with breast cancer, the exploitation of these findings within a practical in vivo test is several years away. However, FCT again offers advantages over conventional X-ray diffraction modes of measurement. Initial results where FCT has been applied to bone phantoms have clearly demonstrated that, at least in transmission, the emerging diffraction signatures are of sufficient quality to enable application of the fracture predictive algorithms, i.e. the magnitude of derived parameter errors are such that fracture and non-fracture groups can be discriminated [59]. Successful commercialisation of this approach holds the promise of portable devices for point-of-patient care thus impacting significantly on the health burden of this condition within an increasingly aging population.

6.4.4 Process Control

The industrial uses of powder diffraction are extensive and include electronics, optics, aerospace, petroleum, mining, cement, paints, pharmaceuticals, forensic, and medicine. The technique is especially useful in the manufacturing industries that require some measures of quality control. For example, the cement industry employs off-line powder diffraction at several points within the manufacturing process to ensure that the various types of cement are being formed with the appropriate chemistry. The pharmaceutical industry uses diffraction extensively for drug discovery and the fabrication of endoprosthetic implant coatings. The semiconductor sector exercises diffraction techniques to ensure the quality of substrates and active thin films, and the aerospace manufacturing exploits diffraction to ensure the attributes of components such as turbine blades. A comprehensive survey of the industrial use of diffraction can be found elsewhere [26].

In general, industries adopt X-ray diffraction due to its unique ability to provide information such as material phase/composition and preferred orientation. A key feature of any analytical method applied within an industrial process is speed of diagnosis and decreases in analysis turnaround time are continuously sought. This is especially the case for high-energy usage industrial processes. For example, it is common practice in the cement industry to sample typically every 2 hours thus, given the rate of production, any unwanted deviation from an ideal composition can result is significant wastage. This situation is compounded by the concomitant energy, fuel, and CO_2 emission waste. Consequently, this change to an in-line, 'real time' approach would be an attractive option but is currently unavailable due to the inherent data collection speeds of X-ray diffraction.

References

1. W. Rontgen, "On a new kind of rays," *Science*, vol. 3, no. 59, pp. 227–231, 1896.
2. W. Friedrich, P. Knipping, M. Laue, "Berichtigung zu der Arbeit: 'Interferenzerscheinungen an Röntgenstrahlen," *Annalen der Physik*, vol. 347, no. 15, pp. 1064–1064, 1913.
3. E. Garman, "Developments in X-ray crystallographic structure determination of biological macromolecules," *Science*, vol. 343, no. 6175, pp. 1102–1108, 2014.
4. W. Bragg, "The reflection of X-rays by crystals," *Proceedings of the Royal Society of London*, vol. 88, no. 605, pp. 428–438, 1913.
5. P. Fratzl, "Small-angle scattering in materials science: a short review of applications in alloys, ceramics and composite materials," *Journal of Applied Crystallography*, vol. 36, pp. 397–404, 2002.
6. B. S. S. Cullity, *Elements of X-Ray Diffraction*, Boston: Prentice Hall, 2001.
7. M. Woolfson, *An Introduction to X-Ray Crystallography*, Cambridge: CUP, 1997.

8. G. Taylor, "The phase problem," *Acta Crystallographica D*, vol. 59, pp. 1881–1890, 2003.
9. J. Watson and F. Crick, "The structure of DNA," in *Cold Spring Harbor Symposia on Quantitative Biology*, 1953.
10. M. Perutz, "Haemoglobin: Structure, function and synthesis," *British Medical Bulletin*, vol. 32, no. 3, pp. 193–194, 1976.
11. D. Chernyshov, "Crystallography with synchrotron light," *Journal of Physics D*, vol. 48, no. 50, 2015.
12. C. Groom and F. Allen, "The Cambridge structural database in retrospect and prospect," *Angewandte Chemie International Edition*, vol. 53, no. 3, pp. 662–671, 2014.
13. T. Okumura, M. Suzuki, H. Nagasawa, T. Kogure, "Microstructural control of calcite via incorporation of intracrystalline organic molecules in shells," *Journal of Crystal Growth*, vol. 381, pp. 114–120, 2013.
14. J. Hanawalt and H. Rinn, "Identification of crystalline materials: Classification and use of X-ray diffraction patterns," *Powder Diffraction*, vol. 1, no. 1, pp. 2–6, 1986.
15. S. N. Kabekkodu, J. Faber and T. Fawcett, "New powder diffraction file (PFD-4) in relational database fromat: Advantages and data-mining capabilities," *Acta Crystallographica Section B*, vol. 58, no. 3, pp. 333–337, 2002.
16. H. Rietveld, "A profile refinement method for nuclear and magnetic structures," *Journal of Applied Crystallography*, vol. 2, pp. 65–71, 1969.
17. R. Young, *The Rietveld Method*, Oxford: OUP, 1993.
18. J. Taylor and L. Alderidge, "Phase analysis of Portland Cement by full profile standardless quantitative X-ray diffraction - accuracy and precision," *Advances in X-Ray Analysis*, vol. 36, pp. 309–314, 1994.
19. H. Klugg and L. Alexander, *X-Ray Diffraction Procedures for Polycrystalline and Amorphous Materials*, London: John Wiley, 1974.
20. A. Narten and H. Levy, "Liquid water: Molecular correlation functions from X-ray diffraction," *Journal of Chemical Physics*, vol. 55, no. 5, pp. 2263–2269, 1971.
21. M. Wilson, L. Dummott, D. Duarte, "A 10cm×10cm CdTe Spectroscopic Imaging Detector based on the HEXITEC ASIC," *Journal of Instrumentation*, vol. 10, pp. 1–13, 2015.
22. L. Hatcher and P. Raithby, "Dynamic single-crystal diffraction studies using synchrotron radiation," *Coordination Chemistry Reviews*, vol. 227, pp. 69–79, 2014.
23. Y. Shvyd'ko, *X-Ray Optics*, Berlin: Springer, 2004.
24. E. Bertin, *Principles and Practice of X-Ray Spectrometric Analysis*, New York: Plenum Press, 1984.
25. R. Dinnebier and S. Billenge, *Powder Diffraction, Theory and Practice*, Cambridge: RSC, 2008.
26. F. Chung and D. Smith, *Industrial Applications of X-Ray Diffraction*, New York: Marcel Dekker, 2000.
27. I. Cotte-Rodríguez, H. Hernández-Soto, H. Chen and R. Cooks, "In situ trace detection of peroxide explosives by desorption electrospray ionization and desorption atmospheric pressure chemical ionization," *Analytical Chemistry*, vol. 80, no. 5, pp. 1512–1519, 2008.
28. A. Jimenez and M. Navas, "Detection of explosives by chemiluminescence," in *Counterterrorism Detection Techniques of Explosives*, Amsterdam: Elsevier, 2007, pp. 1–40.

29. D. Rondeschagen, G. Arnold, S. Bockish, P. Francke, J. Leonhardt and A. Kuster, "Trace and bulk detection of explosives by ion mobility spectrometry and neutron analysis," in *Detection of Liquid Explosives and Flammable Agents in Connection with Terrorism*, Dordrecht: Springer, 2008, pp. 123–133.

30. S. Singh and M. Singh, "Explosives detection systems (ADS) for aviation security: A review," *Signal Processing*, vol. 83, pp. 31–55, 2003.

31. D. Nikolelis; G. Nikoleli, *Biosensors for Security and Bioterrorism Applications*, Switzerland: Springer, 2016.

32. K. K. Z. Whetstone, "A review of conventional explosives detection using active neutron interrogation," *Journal of Radioanalytical and Nuclear Chemistry*, vol. 301, no. 3, pp. 629–639, 2014.

33. L. Cardona, J. Jiménez and N. Vanegas, "Nuclear quadrupole resonance for explosive detection," *Ingeniare. Revista chilena de ingeniería*, vol. 23, no. 3, pp. 458–472, 2015.

34. J. Choi, S. Ryu, W. Kwon, K. Kim and S. Kim, "compound explosives detection and component analysis via terahertz time-domain spectroscopy," *Journal of the Optical Society of Korea*, vol. 17, no. 5, pp. 454–460, 2013.

35. J. Hao, K. Kang, L. Zhang and Z. Chen, "A novel image optimization method for dual-energy computed tomography," *Nuclear Instruments and Methods in Physics Research Section A: Accelerators, Spectrometers, Detectors and Associated Equipment*, vol. 722, pp. 34–42, 2013.

36. G. Harding, "The design of direct tomographic energy-dispersive x-ray diffraction imaging (XDI) systems," in *Proc. SPIE 5923, Penetrating Radiation Systems and Applications VII*, San Diego, 2005.

37. G. Harding, H. Fleckenstein, D. Kosciesza, S. Olesinski, H. Strecker, T. Theedt and G. Zienert, "X-ray diffraction imaging with the multiple inverse fan beam topology: Principles, performance and potential for security screening," *Applied Radiation and Isotopes*, vol. 70, pp. 1228–1237, 2012.

38. K. Rogers, P. Evans, J. Rogers, J. Chan and A. Dicken, "Focal construct geometry - a novel approach to the acquisition of diffraction data," *Journal of Applied Crystallography*, vol. 43, pp. 264–268, 2010.

39. P. Evans, K. Rogers, J. Chan, J. Rogers and A. Dicken, "High intensity X-ray diffraction in transmission mode employing an analog of Poisson's spot," *Applied Physics Letters*, vol. 97, no. 20, 2010, 204101_1-204101_3.

40. K. Rogers, P. Evans, D. Prokopiou, A. Dicken, S. Godber and J. Rogers, "Fundamental parameters approach applied to focal construct geometry for X-ray diffraction," *Nuclear Instruments and Methods in Physics A*, vol. 690, pp. 1–6, 2012.

41. E. Fenimore, "Coded aperture imaging with uniformly redundant arrays," *Applied Optics*, vol. 17, no. 3, pp. 337–347, 1978.

42. J. P. O. Evans, K. D. Rogers, C. Greenwood, A. J. Dicken, S. X. Godber, D. Prokopiou, N. Stone, J. G. Clement, I. Lyburn, R. M. Martin and P. Zioupos, "Energy-dispersive X-ray diffraction using an annular beam," *Optics Express*, vol. 23, no. 10, p. 13443, 2015.

43. F.Li, Z. Liu, T. Sun, B. Jiang and Y. Zhu, "Focal construct geometry for high intensity energy dispersive X-ray diffraction based on X-ray capillary optics," *Journal of Chemical Physics*, vol. 144, no. 10, 2016.

44. J. P. O. Evans, S. X. Godber, F. Elarnaut, D. Downes, A. J. Dicken and K. D. Rogers, "X-ray absorption tomography employing a conical shell beam," *Optics Express*, vol. 24, no. 25, p. 29048, 2016.

45. P. Evans, K. Rogers, A. Dicken, S. Godber and D. Prokopiou, "X-ray diffraction tomography employing an annular beam," *Optics Express*, vol. 22, no. 10, p. 11930, 2014.

46. K. Welles and D. Bradley, "A review of X-ray explosives detection techniques for checked baggage," *Applied Radiation and Isotopes*, vol. 70, no. 8, pp. 1729–1746, 2012.

47. S. Boyd, "Diagnostic applications of high-throughput DNA sequencing," *Annual Review of Pathology: Mechanisms of Disease*, vol. 8, pp. 381–410, 2013.

48. R. Siegel, K. Miller and A. Jemal, "Cancer statistics, 2016," *CA: A Cancer Journal for Clinicians*, vol. 66, no. 1, pp. 7–30, 2016.

49. R. A. Lewis, K. D. Rogers, E. T.-A. C. J. Hall, S. Slawson, A. Evans, S. E. Pinder, I. O. Ellis, C. R. M. Boggis, A. P. Hufton and D. R. Dance, "Breast cancer diagnosis using scattered X-rays," *Journal of Synchrotron Radiation*, vol. 7, pp. 348–352, 2000.

50. R. Scott, N. Stone, C. Kendall, K. Geraki and K. Rogers, "Relationships between pathology and crystal structure in," *npj Breast Cancer*, vol. 2, p. 106029, 2016.

51. R. Baker, K. D. Rogers, N. Shepherd and N. Stone, "New relationships between breast microcalcifications and cancer," *British Journal of Cancer*, vol. 103, pp. 1034–1039, 2010.

52. NOS, "National Osteoporosis Society," 2017. [Online]. Available: https://nos.org.uk/.

53. Z. Syed and A. Khan, "Bone densitometry: Applications and limitations," *Journal of Obstetrics and Gynaecology*, vol. 24, no. 6, pp. 476–484, 2002.

54. T. Sakae, H. Nakada and J. LeGeros, "Historical review of biological apatite crystallography," *Journal of Hard Tissue Biology*, vol. 24, no. 2, pp. 111–122, 2015.

55. S. Toshiro, T. Kono, H. Okada, H. Nakada, H. Ogawa, T. Tsukioka and T. Kaneda, "X-ray micro-diffraction analysis revealed the crystallite size variation in the neighboring regions of a small bone mass," *Journal of Hard Tissue Biology*, vol. 26, no. 1, pp. 103–107, 2017.

56. A. Grunenwalda, C. Keyserb, A. Sautereau, E. Crubézyc, B. Ludesb and C. Droueta, "Revisiting carbonate quantification in apatite (bio)minerals: A validated FTIR methodology," *Journal of archaeological science*, vol. 49, pp. 134–141, 2014.

57. C. Greenwood, J. Clement, A. Dicken, J. Evans, I. Lyburn, R. Martin, K. Rogers, N. Stone and P. Zioupos, "Towards new material biomarkers for fracture risk," *Bone*, vol. 93, pp. 55–63, 2016.

58. A. J. Dicken, J. P. O. Evans, K. D. Rogers, N. Stone, C. Greenwood, S. X. Godber, J. G. Clement, I. D. Lyburn, R. M. Martin and P. Zioupos, "Classification of fracture and non-fracture groups by analysis of coherent X-ray scatter," *Scientific Reports*, vol. 6, p. 29011, 2016.

59. A. J. Dicken, J. P. O. Evans, K. D. Rogers, N. Stone, C. Greenwood, S. X. Godber, D. Prokopiou, J. G. Clement, I. D. Lyburn and R. M. Martin, "X-ray diffraction from bone employing annular and semi-annular beams," *Physics in Medicine and Biology*, vol. 60, no. 15, 2015.

7

X-Ray Diffraction Tomography: Methods and Systems

Shuo Pang and Zheyuan Zhu

CREOL-College of Optics and Photonics

CONTENTS

7.1 Introduction

X-ray computed tomography (CT) is a volumetric imaging technique widely used in medical diagnosis and industrial inspection. Conventional CT captures a series of projections of the object, each of which is the attenuation integrated along a set of X-ray beams, under different projection angles. The back-projected attenuation map is the primary contrast mechanism to distinguish different substances inside the object. Because the X-ray attenuation is mainly related to the electron density, different types of soft tissues, which typically share similar atomic compositions but different structure at molecular level, cannot be well distinguished on transmission-based CT image. For example, in the diagnosis of breast cancer, the difference between healthy and diseased breast tissues involves only subtle structural change, which is not sensitive to conventional attenuation-based X-ray mammography [1]. Therefore, additional histopathology examination might be required for accurate diagnosis [2]. The X-ray coherent scattering signal, on the other hand, has proven to be significantly stronger and different in the cancerous regions compared to the healthy region [3], due to the change of axial period of collagen fibrils. In security screening applications, conventional transmission-based X-ray image can only identify suspicious objects based on their structural profile. Coherent scattering can discriminate explosives from non-hazardous items based on the location of scattering peaks, which are considered as fingerprints unique to explosives [4,5]. In both medical and security screening applications, X-ray coherent scattering can improve material specificity without contrast agent.

For the coherent scattering imaging of an extended object, the scattering profile of each point can be reconstructed by combining X-ray diffraction measurement with a projection setup similar to conventional CT. This method is termed as X-ray diffraction tomography (XDT), which has shown its success in resolving material-specific diffraction profile within the sample [6–8] and has also been demonstrated feasible in medical imaging settings [9,10]. XDT setups generally fall into three categories: direct tomography that filters out one particular scattering direction using collimators in front of the detector [11,12], angular-dispersive tomography using a narrow-band source and energy-integrating detector arrays [8,13], and energy-dispersive tomography using a broad-band source and energy-sensitive detectors [14,15].

All these XDT implementations assumes the material is amorphous or in the powder form, which produces isotropic scattering profiles that only depends on the angular difference between incident and diffracted X-ray, rather than the three-dimensional diffraction direction. However, this does not apply to all the biological tissues; bones/teeth [16,17], and many biocompatible materials used for medical implants [18] all possess partial long-range ordering at molecular level. Many explosive substances also contain crystalline structures that give rise to anisotropic scattering signatures [19]. The isotropic scattering assumption in conventional XDT could falsify the image reconstruction when scattering textures are present.

Another problem is the weak intensity of the diffracted X-ray compared to that of the transmitted beam. As a result, many scattering signatures mentioned above are measured with high-intensity a synchrotron source, which limits their applications in practice. A high-resolution X-ray scattering imaging system using a tabletop X-ray tube would accelerate the adoption of this method for clinical and security use [20]. However, the use of a low-brilliance tabletop X-ray tube requires long integration time to overcome the noise on the detector, leading to lengthy total acquisition time, and high-dosage exposure to ionized radiation. Moreover, in XDT systems, it is inevitable to use either a source-side collimator to localize the scattering volume or a detector-side collimator to limit the scattering angle.

The use of collimators throws away a large portion of diffracted X-ray photons, resulting in low-collection efficiency. When a high-resolution XDT image is desired, the acquisition time becomes longer as better localization is required. Therefore, there is a trade-off between resolution and acquisition time in conventional XDT systems.

This chapter first reviews existing X-ray diffraction imaging modalities, including direct tomography, pencil beam angular dispersive tomography, fan beam angular dispersive tomography, and energy-dispersive tomography. Then we address the vector momentum space reconstruction, which addresses the issue of textured sample. In brief, to recover the full three-dimensional scattering distribution within a two-dimensional sample layer, we increase the dimension of our data acquisition to match that of the object function by introducing the movement of the detector along the beam. We also introduce a localized reconstruction scheme to reduce the acquisition time and radiation dose for XDT. Only a region-of-interest (ROI) inside the object is scanned, making XDT an appealing modality for secondary high-specific imaging applications.

7.2 Principles of Coherent Scattering Tomography

7.2.1 X-Ray Diffraction from a Single Voxel

To simplify the discussion, we assume the X-ray tube focus is far from the sample and the illumination thus can be treated as a parallel beam geometry. The diffracted photon count, $dI(E)$, from a scatter voxel dV at scattering angle θ is:

$$dI(E,\theta) = I_0(E)dVn_0\frac{d\sigma(E,\theta)}{d\Omega}, \tag{7.1}$$

where $I_0(E)$ is the incident X-ray photon number per cm², n_0 is the number of scatters per cm³, and $d\sigma/d\Omega$ is the differential cross-section of diffraction, which has a unit of cm⁻² per steradian, and

$$\frac{d\sigma}{d\Omega} = \frac{r_e^2}{2}\left(1+\cos^2\theta\right)f_0(q,\mathbf{r}), \tag{7.2}$$

where r_e is the classical electron radius, θ is the scatter angle, $f_0(q, \mathbf{r})$ is the molecular form factor at location $\mathbf{r} = (x, y, z)$, and q is the momentum transfer. According to the Bragg's law:

$$q = \frac{E\sin\theta/2}{hc}, \tag{7.3}$$

where h is the Planck constant and c is the speed of light. The product of the scatter density and the form factor, $f(q, \mathbf{r}) = n_0(\mathbf{r})f_0(q, \mathbf{r})$, is what we call the object function and want to reconstruct. The geometry is shown in Figure 7.1(a). The profile of the object function indicates the type of material and the amplitude indicates the density. Combining Eq. (7.1)–(7.3), the detected scatter irradiance on the detector located at $\mathbf{r}_d = (x_d, y_d, z_d)$ is

$$I_{\mathrm{coh}}(E, \mathbf{r}_d) = I_0(E) \int_V \int_q \int_\Omega \frac{r_e^2}{2}(1 + \cos^2 \theta)\delta\left(q - \frac{E \sin \theta/2}{hc}\right)f(q, \mathbf{r})\,d\Omega\,dq\,dV, \quad (7.4)$$

where Ω is the collection solid angle covered by the detector.

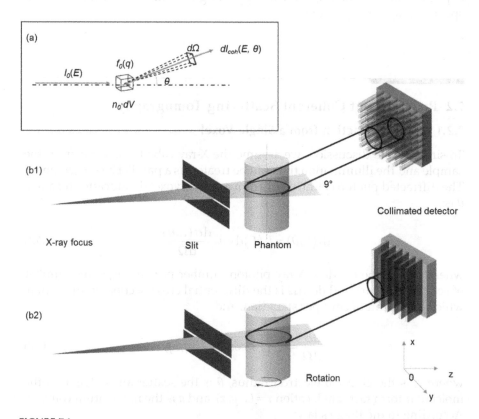

FIGURE 7.1
(a) Geometry of coherent scattering. (b) The schematics of coherent scatter imaging setups: direct tomography with polycapillary collimators (b1), and fan-beam X-ray diffraction tomography (XDT) with collimation only in the horizontal direction (b2).

7.2.2 Direct Tomography

Equation (7.4) shows that given the energy of the X-ray photon, each detector pixel collects the scatter photons from every object voxel at a particular scatter angle. A simple means to distinguish the scatter photons from different voxels is the use of a collimator. Figure 7.1(b1) shows the system setup, where each detector pixel only receives collimated photons from a small object volume. The system is termed as *direct tomography*.

Specifically, a 2D detector array with capillary collimators is angled by θ. The collection cone of each detector pixel has an intersection area with the illumination plane. Let Ω be the collection solid angle of the detector. The direction cosine of the normal direction of the detector pixel is $p = (a, b, c)$, where $a^2 + b^2 + c^2 = 1$. Since the incident beam is parallel, the scattered ray vector is $r_{sc} = (x'-x, y'-y, z'-z)$. The angular cosine between the scatter ray and the normal direction of detector is $\cos\gamma = \dfrac{r_{sc} \cdot p}{\|r_{sc}\| \|p\|}$. The collimator, limiting the acceptance angle, can be modeled as a rectangle function of γ.

$$t(\gamma) = \mathrm{rect}\left(\frac{\gamma}{\beta}\right) = \begin{cases} 1, \ |\gamma| \leq \beta \\ 0, \ |\gamma| > \beta \end{cases}, \tag{7.5}$$

where β is the angular range of polycapillary tubes, which is around 1 mrad [21]. Combining Eqs. (7.4) and (7.5), the collected scatter photon counts at energy E is

$$g_0(E, r_d) = \iint I_0(E) K(r, r_d) f(q, r) \delta\left(q - \frac{E \sin\theta/2}{hc}\right) dq \, dV, \tag{7.6}$$

where $K(r, r_d)$ is the blur kernel.

$$K(r, r_d) = \frac{r_e^2}{2} \int\limits_{y_d - \Delta/2}^{y_d + \Delta/2} \int\limits_{x_d - \Delta/2}^{x_d + \Delta/2} (1 + \cos^2\theta) \frac{\cos\gamma}{\|r - r'\|^2} t(\gamma) dx' dy', \tag{7.7}$$

where r' is the convolution variable in the detector coordinates and Δ is the pixel size of the detector. Equation (7.7) shows that given a detector to voxel distance, the larger the angular range β, the broader the blur kernel.

Next, considering the energy response of the system, $S(E, E')$, which is the convolution of the spectral response of the detector and the spectrum of the source, we have

$$g(E, r_d) = \int\limits_{E'} g_0(E', r_d) S(E, E') dE', \tag{7.8}$$

where E' is the convolution variable in the energy domain. For an ideal energy-integrating detector array and monochromatic source centered at E_0, which are used in angular dispersive setups, the energy convolution kernel can be treated as a Dirac-delta function. Then the measurement can be simplified as

$$g(\mathbf{r}_d) = I_0(E_0) \iint K(\mathbf{r}, \mathbf{r}_d) f\left(\frac{E_0 \sin\theta/2}{hc}, \mathbf{r}\right) dV. \tag{7.9}$$

Equation (7.9) shows that we can measure the form factor f at various momentum transfer q by either changing the source energy E_0 or scatter angle θ [15,21].

7.2.3 Pencil Beam Angular Dispersive X-Ray Diffraction Tomography (AD-XDT)

The use of polycapillary collimators would directly localize sources of scatter signals, unfortunately, at the expense of the collection efficiency. Also, due to the small scatter angle θ, the resolution along the beam illumination direction is lower than that along the perpendicular direction. By rotating the sample in the illumination plane and performing reconstruction similar to conventional CT, an improved resolution can be achieved along the illumination direction. Similar to the first-generation CT, pencil beam angular dispersive XDT uses a pencil beam to probe the sample, and a one-dimensional or two-dimensional array collects the diffracted photons. The major advantage of using pencil beam is that the collimation can be completely eliminated, making the highest collection efficiency possible. The pencil beam setup is ideal for observing samples with small scatter cross section. To achieve the resolution similar to the detector pixel size, the pencil beam has to have a diameter on the order of $0.1 \sim 1$ mm, which would greatly reduce the efficiency of the source [22,23].

For the pencil beam setup, the diffraction process is same as that of a single voxel. The horizontal resolution is determined by the width of the pencil beam. According to Eqs. (7.6) and (7.9), the measurement of one pencil beam by the energy-sensitive detector is:

$$g(s, E, w) = \int I_0(E) f(s, z, q) \delta\left(q - \frac{E \sin\theta/2}{hc}\right) dz\,dq, \tag{7.10}$$

where s is the horizontal offset of the pencil beam in this case and w is the radial offset of the detector. The scatter angle $\theta = r/(l-z)$. Plug the scatter angle into Bragg's condition, we get:

$$qz = \frac{Er}{2lhc}. \tag{7.11}$$

Equation (7.11) shows that the product, Er, is the scaling parameter of the hyperbola. In other words, the E and r are coupled for each measurement. In order to get the range information, we need to introduce both the translational and rotational movement to the sample and get a four dimensional measurement $g(s,\phi,E,r)$. Since E and r are coupled, we are introducing parameter $\eta = Er$:

$$g(s,\phi,\eta) = \int g\left(s,\phi,\frac{\eta}{r},r\right)\mathrm{d}r. \tag{7.11}$$

The projection is a mapping from 3D object space (x,y,q) to 3D measurement space (s,ϕ,η), equivalent to a parallel beam 3D CT reconstruction problem [24].

7.2.4 Fan Beam Angular Dispersive X-Ray Diffraction Tomography

However, the source utilization efficiency of the pencil beam system is extremely low. Fan-beam XDT systems illuminate a plane of the sample (c.f. Figure 7.1(b2)) and performing reconstruction similar to the pencil beam setup. The fan beam AD-XDT geometry is also termed as coherent scatter computerized tomography (CSCT) [13,25,26]. The detectors in fan beam AD-XDT setup only needs the collimation in horizontal direction and collects diffracted photons from all voxels along the illumination direction. The setup using a 2D energy-integrating detector array was described in [8].

The cross-sectional view of the detection geometry is shown in Figure 7.2(a1). In this setup, the energy-integrating detector rows are placed at a distance, w, away from the illumination plane similar to the pencil beam AD-XDT. At each projection angle, the measurement is a 2D dataset (y_d along the detector row and w along the vertical direction). Similar to transmission CT, a sequence of 2D projections are measured while the object is being rotated.

Assuming that the object dimension, Δl, is much smaller than the distance from the object to the detector, l_0, (e.g., $\Delta l/l_0 < 0.05$), and the scatter angle is in the small angle regime (i.e., $\theta_0 \approx \tan\theta_0 = \frac{w}{l_0}$, and $\sin\theta \approx \frac{w}{l_0 - \Delta l}$), we can simplify the forward model. The scatter angle change, $\Delta\theta$, due to the small displacement of the scatter point from the object center, Δl, can be approximated as $\Delta\theta \approx \frac{\Delta l}{l_0}\theta_0$. The detector is tilted so that it is perpendicular to the scatter ray from the center of the object, $\gamma = \frac{\theta_0(l_0 - \Delta l) - w}{l_0}$ and $\cos = 1 - \frac{1}{2}\gamma^2 + \cdots \approx 1$.

Because the detector pixel size, Δ, is much smaller than the distance l, we can assume the irradiance is uniform at each pixel. Since no collimation in the horizontal direction, $t(\gamma) = 1$. As a result, the spatial convolution kernel in Eq. (7.7) can be simplified to

$$K(\mathbf{r},\mathbf{r_d}) \approx \frac{\Delta^2 r_e^2}{2}\left(1+\cos^2\theta_0\right)\left[\frac{1}{\left(l_0-\Delta l\right)^2+w^2}-\frac{\theta_0\sin 2\theta_0}{{l_0}^2+w^2}\frac{\Delta l}{l_0}\right]. \quad (7.12)$$

Since $\Delta l/l_0$ and θ_0 are small, we can drop the second term in Eq. (7.10), which is only ~0.2% of the first term. Plugging the first term of Eq. (7.12) to Eq. (7.6), we have

$$g_0(w,\mathbf{r_d}) = \frac{\Delta^2 r_e^2}{2}\left(1+\cos^2\theta_0\right)I_0(E_0)\int_{\mathbf{r}}\frac{1}{\left(l_0-\Delta l\right)^2+w^2}f\left(\frac{w}{2hc}\frac{E_0}{l_0+\Delta l},\mathbf{r}-\mathbf{r_d}\right)dV$$

$$(7.13)$$

Similar to Eq. (7.9), Eq. (7.13) assumes monochromatic source has an ideal energy profile centered at E_0. The projection transform described in Eq. (7.13) can be treated as a line integration on the q–Δl plane along the family of curves:

$$q = \frac{w}{2hc\left(l_0-\Delta l\right)}E_0, \quad (7.14)$$

with w as the parameter, shown in Figure 7.2(a2). The 3D tomography reconstruction can be carried out similar to the case of attenuation CT. The back

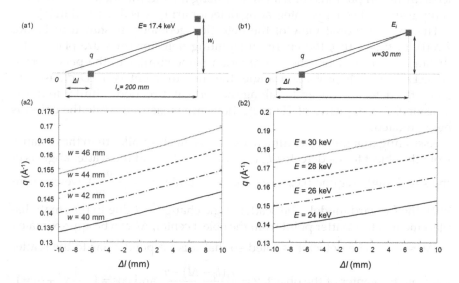

FIGURE 7.2

(a) Angular dispersive measurement geometry: side view of AD-XDT geometry (a1); each row of the detection at different w is a line integration along a section of a hyperbolic curve in q–Δl space (a2). (b) Energy-dispersive measurement geometry: side view of ED-XDT geometry (b1); the detection for each energy channel is also a line integration along a specific hyperbolic curve in q–Δl space (b2).

projection is from the measurement domain of (w, y_d, ϕ) to the object domain of (x, y, q), where ϕ is the rotation angle.

7.2.5 Energy-Dispersive Coherent Scatter Computerized Tomography

Energy-dispersive X-ray diffraction tomography (ED-XDT) employs a broadband source without the need for filtration and thus can improve the x-ray source efficiency by orders of magnitude. The momentum transfer resolution relies on the energy resolution of the system. In AD-XDT, the normalized energy resolution is the ratio between the peak-width and the peak of the characteristic line of the source, which is on the order of 10%. Until recently, the normalized energy resolution of array detectors, σ_E/E, was worse than 10% [26]. In addition, the energy-discriminative detector usually has a larger pixel size than energy-integrating detector, resulting in an inferior resolution [22]. The recent advancement of CdTe/CZT photon-counting detector array has improved the energy resolution to less than 6 keV for hard X-ray, with the detector pixel size as small as 55 μm [27]. The application of such detectors in ED-XDT could improve the system resolution, while providing a faster acquisition time than AD-XDT system.

In AD-XDT, energy-integrating detector array measures the single energy diffraction at different diffraction angles at corresponding vertical offset, w, while each detector row in ED-XDT measures the broadband diffracted photons in corresponding energy channels. As a result, the measurement dataset is in the domain of (E, y_d, ϕ), rather than (w, y_d, ϕ). The system geometry is similar to that of the AD-XDT, and the projection onto a particular energy channel centered at E_i is also carried out on the $q-\Delta l$ plane along the curve described by Eq. (7.14), but with energy E as the parameter this time, as shown in Figure 7.2(b).

Here we assume the spectrum of the detector has a Gaussian function centered at E, with energy deviation of σ_E. Then the energy convolution kernel can be expressed as

$$S(E - E') = \frac{1}{\sqrt{2\pi}\sigma_E} \exp\left(-\frac{(E-E')^2}{2\sigma_E{}^2}\right). \tag{7.15}$$

For the energy channel centered at E_i, the measurement is

$$g(E_i, \mathbf{r}_d) = \frac{\Delta^2 r_e{}^2}{2}\left(1 + \cos^2\theta_0\right)I_0(E_i)\int_{\mathbf{r}}\int_{q}\frac{f(q, \mathbf{r} - \mathbf{r}_d)}{(l_0 + \Delta l)^2 + w_0{}^2}S\left(E_i - \frac{2hc}{q}\frac{l_0 - \Delta l}{w_0}\right)dq\,dV, \tag{7.16}$$

The momentum transfer resolution for both AD-XDT and ED-XDT relies on the variance of the energy distribution of the system. In AD-XDT, the spectrum broadening of the source deteriorates the momentum transfer resolution. The normalized energy resolution is the ratio between peak width and the characteristic emission line of the X-ray source, which is typically around 10%. A typical energy-sensitive detector array, on the other hand, has a normalized energy resolution worse than 10% until recently. In addition, the energy-discriminative detector usually has a larger pixel size than the energy-integrating detector, resulting in an inferior spatial resolution [22]. We have compared the reconstruction performance of ED-XDT using an energy-sensitive detector with a Gaussian energy response spectrum and AD-XDT using a realistic copper-anode tube spectrum. The AD-XDT and ED-XDT measurements were simulated using a phantom consisting of normal and cancerous breast tissues, as shown in Figure 7.3, and reconstructed to obtain the coherent scattering profile of each tissue type. Figure 7.3(c) shows the ED-XDT reconstruction error of normal and carcinoma tissue as a function of energy resolution of the energy-sensitive detector, along with the AD-XDT reconstruction error under copper and silver-anode X-ray tube. The reconstruction from the silver-anode tube measurement exhibit larger error compared to copper-anode tube due to its broader spectrum. We have found that with the development of energy-sensitive detector array, the ED-XDT could reach a comparable performance as copper-anode AD-XDT scheme. Specifically, if the energy resolution of the detector is better than 7 keV (~10% normalized energy resolution), the reconstruction from ED-XDT measurement could outperform AD-XDT scheme that typically uses a copper-anode tube [14].

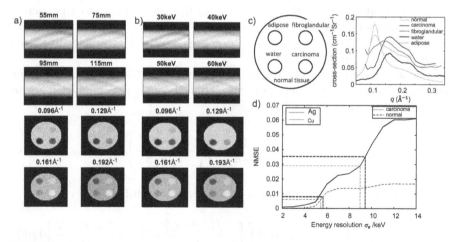

FIGURE 7.3
Comparison between AD-XDT and ED-XDT measurement and reconstruction. (a) AD-XDT sinograms under w = 55, 75, 95, and 115 mm, and reconstructed spatial profiles under q = 0.096, 0.129, 0.161, and 0.192 Å^{-1}. (b) ED-XDT sinograms under 30, 40, 50, and 60 keV energy channels, and the reconstructed spatial profiles under the same q in (a). (c) XDT phantom and the scattering profile of each tissue used in the simulation. (d) Reconstruction error of ED-XDT for carcinoma and normal tissue under different energy resolution of the detector array.

7.3 Recovering Full-Momentum Transfer Space Profile

The aforementioned XDT systems can only reconstruct the scattering profile as a function of the magnitude of momentum transfer vector q_r. To reconstruct the full three-dimensional diffraction distribution within a two-dimensional sample layer $F(x, y = 0, z, q_x, q_y, q_z)$, a total of five dimensions need to be supplied in the measurement to match the dimension of the object function [28]. We used a copper-anode X-ray tube, which emits a narrow peak around 8 keV (wavelength = 1.5 Å), along with an energy-integrating panel detector. The source spectrum $I_0(E)$ in Eq. (6) can be treated as a δ-function and does not contribute to additional dimension in the measurement. Previous attempts to preserve the direction-dependent scattering textures captured a series of 2D small-angle X-ray scattering (SAXS) images but only managed to reconstruct the spatial profile along the rotation-invariant scattering direction [28–30], leaving a large portion of the collected data unused. Recently, Schaff *et al.* added the tilt of rotational axis to the data acquisition scheme and demonstrated the reconstruction of full three-dimensional real and reciprocal space of the sample [23]. However, the use of a synchrotron source and a complicated rotation motion limit its application in practical settings. Moreover, for imaging a two-dimensional slice within a three-dimensional object, or a thin slice of flat object, this data acquisition scheme suffers from dimension insufficiency, since tilting the rotational axis and raster scan perpendicular to the sample slice does not capture more information about the slice of interest.

To solve the texturing problem for a 2D sample layer, we have introduced the detector movement along z_d to supply the dimensions needed for the five-dimensional reconstruction of the object function. The resultant measurement on the detector under different position z_d is

$$g_{s,\phi}(w, z_d, \psi) =$$

$$\iint I_0 r_e^2 F(\mathbf{A}_\phi(s,0,t)^T, \mathbf{A}_\phi \mathbf{q}) \delta \left[\mathbf{q} - \frac{E}{hc} \left(\frac{w \hat{r}}{z_d - t} \right) \right] \frac{\Delta^2}{w^2 + (z_d - t)^2} A \, dt \, d\mathbf{q}, \quad (7.17)$$

The five-dimensional data acquired as a result of sample shift, rotation, and detector translation is illustrated in Figure 7.4, which is a series of four-dimensional sinograms along z_d. Each pixel on the 4D sinogram contains an entire 2D scattering image $g_{s,\phi}(w, \psi)$ rather than a single value as the attenuation-based CT.

For a fixed detector location, z_d, and $\psi = 90°$ on the detector plane, the unit vector $\hat{r}$ in Eq. (17) becomes $\hat{y}$, and therefore only the momentum transfer vector $\mathbf{q} = q_y \hat{y}$ is captured by the pencil-beam system. Since the vector $\hat{y}$ is invariant to rotation matrix $\mathbf{A}_\phi$ under all projection angles, the same momentum transfer q_y is probed for all the object voxels as the sample rotates, which

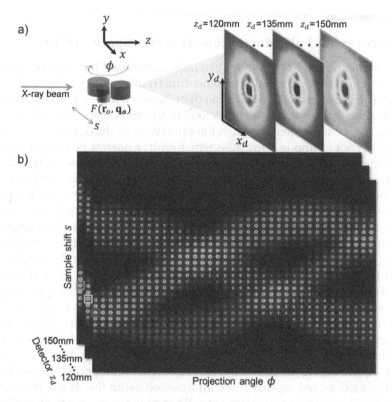

FIGURE 7.4
(a) Measurement scheme of the three-dimensional momentum transfer space inside an extended object, consisting of sample translation and rotation in addition to the pencil-beam coherent scattering model. (b) The five-dimensional measurement consists of a series of 4D sinograms along the z direction. Each pixel in the sinogram under a specific z_d consists of an entire 2D scattering image. The red, yellow, and green squares on the sinogram correspond to the scattering images enclosed in red, yellow, and green in (a).

simplifies the intensity along this direction on the detector to a conventional AD-XDT measurement.

$$g_{s,\phi}(w) = r_e^2 \frac{\Delta^2}{w^2 + (z_d - t)^2} AI_0$$

$$\int F\left(s\cos\phi + t\sin\phi, 0, -s\sin\phi + t\cos\phi, \frac{E}{hc}\left(\frac{w}{z_d - t}\right)\hat{y}\right)dt. \quad (7.18)$$

The projection transform in Eq. (18) can be treated as a line integral on the $q_y - t$ plane along a family of curves $q_y = \dfrac{E}{hc}\dfrac{w}{z_d - t}$ with w as a parameter. The tomographic reconstruction on a subset of the object function $F(x, 0, z, 0, q_y, 0)$

can then be implemented using the AD-XDT scheme described above, which establishes a back projection from the measurement domain (s, ϕ, w) to the object domain (x, z, q_y). Figure 7.5 summarizes the AD-XDT reconstruction result along the rotation-invariant direction y at a spatial resolution of 1 mm (which is limited by the collimator size) and momentum transfer resolution of 0.005 Å^{-1}. Figure 7.5(a) is a normalized transmission map from a CT scan of the sample. The spatial profiles of the sample reconstructed from the intensity along the vertical radial direction on the detector are shown in Figure 7.5(b1)–(b3), corresponding to $q_y = 0.08$, 0.12, and 0.16 Å^{-1}, respectively. The difference in the contrast between water and butter can be easily observed under different momentum transfer values, due to the difference in their scattering peaks. The reconstructed scattering profiles of three pixels marked in green, blue, and red, corresponding to butter, water, and Teflon, are compared to their reference profiles in Figure 7.5(c1)–(c3), respectively.

Except for the rotation-invariant radial direction $\psi = 90°$ on the detector, all other radial directions cannot be directly mapped back to the q_r under the corresponding azimuthal angle using conventional AD-XDT reconstruction, since different 2D momentum transfer sections are probed as projection angle changes. Instead, recovering the scattering profile for all voxels along the beam $f(t, q_r, \psi)$ relies on the inversion of the pencil-beam model described by Eq. (4). However, due to the small scattering angle and limited range of detector movement, voxels along the beam are blurred together on the detector, making it hard to sort out the scattering direction from different positions along the beam. We can solve this problem with the help from the

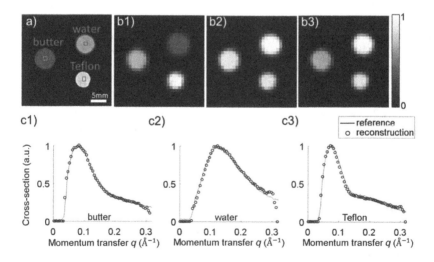

FIGURE 7.5
(a) Transmission CT image of the phantom. (b1–b3) The spatial profile under momentum transfer values of 0.08, 0.12, and 0.16^{-1} along the vertical direction. (c1–c3) The scattering profiles of the pixels marked by green, blue, and red in (a).

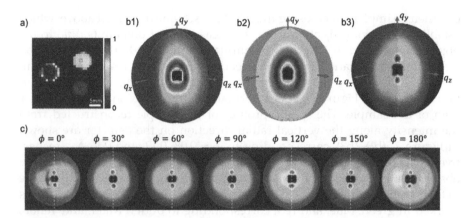

FIGURE 7.6
(a) The spatial profile of the sample under $q_x = 0.08^{-1}$. (b1)–(b3) Two orthogonal sections in the three-dimensional diffraction volume of the pixel marked in green, blue, and red, corresponding to the material of butter, water, and Teflon wrap. (c) The object volume of Teflon is expanded into seven slices under projection angles of 0° to 180° at 30° step. The white, dashed line indicates the y direction to facilitate the display of the orientation of scattering peak as the sample undergoes rotation. The display window of the momentum transfer space is 0–0.3^{-1}.

spatial information obtained in AD-XDT reconstruction. First, the support of the object was acquired by thresholding the AD-XDT reconstruction. (The background is less than 10% of the maximum intensity.) Then $f(t, q_r)$ at every azimuthal angle was reconstructed, with the object support in dimension t as prior information. Finally, the complete five-dimensional function in sample coordinates was recovered through coordinate transform, according to Eq. (5). Our five-dimensional imaging modality may find potential biomedical applications including monitoring the tissue regeneration in porous media [31,32] and assessing the bone health via its different compositions [25]. Partially textured XRD profile could also be used as contrast in security screening applications [19]. The full reciprocal-space coherent scattering tomography scheme could serve as a stand-alone, high-contrast imaging modality in medical imaging, or a secondary screening method in addition to conventional AD-XDT in case anisotropic diffraction signal is present (Figure 7.6).

7.4 Acquisition Time and Dosage Reduction via Interior Tomography

In conventional AD-XDT, the low collection efficiency associated with the use of collimator, along with the intrinsically small scattering cross-section of most amorphous materials, both contribute to the weak scattering signal, especially when a tabletop source is used. Thus, a tabletop XDT setup typically demands a long integration time to overcome the noise on the detector, resulting in a lengthy acquisition when the full field-of-view (FOV) of the object is scanned. However, in many applications, material discrimination only within a small ROI is desired. Tomographic scan focusing on the ROI region (termed "interior tomography") could significantly reduce the imaging time and radiation dosage delivered to the sample. One biggest problem with interior tomography is the insufficient amount of data to uniquely determine the solution inside ROI. It has been proved that in transmission-based CT, exact solution of an interior ROI only from the projection data associated with the beams through interior region is impossible [33], unless certain prior information of the sample is given. For example, under differentiated backprojection framework [34], the inversion of truncated Hilbert transform of an analytically continuous object is made possible by knowing a sub-region inside the ROI [35,36]. Another commonly used prior is the assumption of piecewise constant within ROI, which can be implemented by minimizing the total variation (TV) or high-order TV [37,38]. In the case of XDT, the portion of diffracted signal on the detector attributed to the exterior region outside ROI is unknown and therefore, resolving the spatially variant scattering profile in ROI from limited projection data remains a challenge. Similar to the interior tomography for CT, we have demonstrated that using a TV constrain for AD-XDT reconstruction can preserve the material specificity in ROI region while reducing the integration time and radiation dosage to ~20%. The measurement is similar to AD-XDT, but with truncation along the dimension of sample translation s:

$$g_{s,\phi}(w) = r_e^2 \frac{\Delta^2}{w^2 + (z_d - t)^2} AI_0$$

$$\int F\left(s\cos\phi + t\sin\phi, 0, -s\sin\phi + t\cos\phi, \frac{E}{hc}\left(\frac{w}{z_d - t}\right)\right) dt. \qquad (7.19)$$

For the interior tomographic measurement in Eq. (19), the sample translation s is calculated under each projection angle to cover the ROI region only. Since the scattering photons originate from both interior and exterior region, the full size of the object, rather than the interior region, was discretized and built into the system model, which establishes a linear mapping, $\mathbf{H}$, between the whole object and the truncated measurement $\mathbf{g} = \mathbf{Hf}$. The extended size of the forward model beyond the ROI region allows us to attribute the scattering signals that cannot fit in the priors in the interior region to the pixels outside ROI during the reconstruction. From the measurement $\mathbf{g}$, the object $\mathbf{f}$ is reconstructed by a maximum-likelihood estimator [39] with TV regularization [40,41]

$$\hat{\mathbf{f}} = \arg\min_{\mathbf{f}'}(-\log P(\mathbf{g}\,|\,\mathbf{f}') + \tau TV(\mathbf{f}')), \tag{7.20}$$

where $P(\mathbf{g}\,|\,\mathbf{f}')$ is the Poisson likelihood of observing the measurement $\mathbf{g}$ given the parameters in the vector $\mathbf{f}'$; τ is the weight for balancing the measurement error and the TV regularizer.

The result of an interior XDT experiment to resolve the scattering profile in physical phantom with limited, off-centered projection data. Prior to the experiment, a CT scan was conducted on the actual phantom to capture its general morphology, shown in Figure 7.7(a). A full FOV pencil beam XDT scan was carried out and subsequent data truncation was performed to remove the projection data outside ROI. Since the ROI region, marked by the dashed circle in Figure 7.7(a), is off-centered, the truncation along the beam position dimension s under each projection angle needs to be rounded up to the nearest projection data. Both the complete and truncated measurements were reconstructed to obtain the full FOV and interior XDT image. The difference inside ROI region between full-FOV XDT and interior XDT reconstruction, quantified by the NMSE, is 0.9%. Figure 7.7(c) shows the reconstructed scattering profile $f(q)$ from the truncated measurement compared to the reference profiles in the red, green, and blue squares in (b),

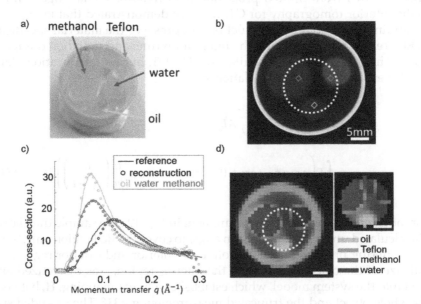

FIGURE 7.7
XDT and interior XDT reconstruction on the phantom. (a) A photo of the sample; (b) normalized attenuation map of the phantom from CT scan. (c) Reconstructed scattering profile from interior XDT measurement in the region marked by red, green, and blue squares in (b), corresponding to the material of methanol, vegetable oil, and water, respectively. (d) Material map for the full FOV and interior XDT. All scale bars represent 5 mm.

corresponding to the region of methanol, vegetable oil, and water in the phantom, respectively. Notice that on the CT image, methanol and oil share almost the same contrast, but their scattering intensity and peak position differ. By comparing the reconstructed scattering profile of each pixel with the reference profile collected from each individual material, a material map (shown in Figure 7.7(d)) can be generated using support vector machine (SVM) [42]. Compared with the full-FOV X-ray tomography scan, the reconstructed scattering profiles of the interior region retain the information necessary to accurately identify the material, while reducing the total acquisition time by three times.

To demonstrate the capability of providing contrasts among biological tissues, we imaged a small piece of meat that contains both fat and muscle. The meat was cut into a 1-inch by 1-inch square from a large piece of smoked ham and air-dried before the experiment to preserve the soft tissue from deformation. Similar to the phantom experiment, a full-FOV XDT measurement was first performed and then truncated to a 6 mm-wide circular ROI region on the upper-right corner as an interior XDT measurement. Figure 7.8(a) shows a CT image of the ham sample. The inset shows two spatial profiles under $q = 0.08$ and $q = 0.16$ Å^{-1}. Fat tissue appears darker compared to muscle because it is a weaker attenuator but is a stronger scatter under low momentum transfer values. The contrast between fat and muscle reverses at high momentum transfer due to the difference in their scattering peak. Figure 7.8(b) displays the reconstructed diffraction profile of the square and diamond marked on the CT image, corresponding to fat and muscle in the ROI, from both the truncated measurement and a full-FOV measurement. The difference between the scattering profile reconstructed from interior and full-FOV XDT measurement, evaluated

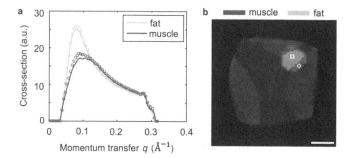

FIGURE 7.8
Interior XDT reconstruction on a piece of ham. (a) Reconstructed coherent scattering cross section of fat and muscle inside the ham sample. The continuous line represents full-FOV XDT reconstruction and the dot represents interior XDT reconstruction. (b) Interior XDT reconstruction with fat and muscle region identified by comparing the scattering profile of each pixel to that in (a). The background is a CT image of the same sample for reference. The scale bar represents 5 mm.

by the NMSE, is 5.4×10^{-4} for fat and 2.9×10^{-3} for muscle. We speculate that the higher discrepancy in the scattering profiles of muscle is caused by its proximity to the ROI boundary, where a deteriorated reconstruction result is commonly observed in interior tomography due to the increasing ambiguity toward the boundary [33]. Nevertheless, the scattering profile of muscle well resembles that of water, while fat gives a strong scatter peak under lower momentum transfer due to the long chains of lipid mono-layer, consistent with the previously reported cross section for biological tissues [6,9]. Using the SVM classification, a material identification was performed for the reconstructed scattering profile inside the ROI. Six scattering profiles representing the fat and muscle outside ROI were selected as training set for two binary SVMs. Each binary SVM was then applied to evaluate the ROI region reconstructed from truncated projection data, and the higher score determines the type of tissue within each 1 mm $\times$ 1 mm pixel. The resulting material map of the ROI is superimposed onto the CT image in Figure 7.8(a), with fat rendered in green and muscle rendered in red. A CT image of the ham sample is shown in the background for reference.

7.5 Conclusions

In summary, X-ray diffraction provides material-specific fingerprint for medical diagnosis and anomaly detection. Combining XRD with tomographic data acquisition, XDT resolves the scattering profile of each point within an extended object, either using the angular-dispersive or energy-dispersive schemes. We have found out that with the development energy-sensitive detector technology, ED-XDT with an energy-sensitive detector array could outperform AD-XDT using a narrow-band X-ray source as long as the energy resolution of the detector ΔE is less than ~5 keV. The scattering texture reconstruction error can be corrected by modeling the full reciprocal space and incorporate additional degrees of freedom into the data acquisition. In addition, interior reconstruction technique can be applied in XDT system, reducing the total acquisition time and radiation damage to the object of interest. Another asset of XDT system is the possibility to combine simultaneous X-ray diffraction with fluorescence and absorption [43]. The demonstration illustrates the capability of XDT and allows us to extract information that is more conclusive in terms of the classification of chemical composition. Similar to CT development, further parallelization in acquisition process would efficiently improve the acquisition time. We envision that such approaches will supersede the more traditional scanning methods and enable more exciting applications in industrial and medical imaging.

References

1. E. A. Ryan and M. J. Farquharson, "Breast tissue classification using X-ray scattering measurements and multivariate data analysis," *Phys. Med. Biol.*, vol. 52, no. 22, pp. 6679–6696, 2007.

2. W. M. Elshemey, F. S. Mohamed, and I. M. Khater, "X-ray scattering for the characterization of lyophilized breast tissue samples," *Radiat. Phys. Chem.*, vol. 90, pp. 67–72, 2013.

3. M. Fernández, J. Keyriläinen, R. Serimaa, M. Torkkeli, M. L. Karjalainen-Lindsberg, M. Tenhunen, W. Thomlinson, V. Urban, and P. Suortti, "Small-angle X-ray scattering studies of human breast tissue samples," *Phys. Med. Biol.*, vol. 47, no. 4, pp. 577–592, 2002.

4. R. W. Madden, J. Mahdavieh, R. C. Smith, and R. Subramanian, "An explosives detection system for airline security using coherent X-ray scattering technology," *Proc. SPIE 7079, Hard X-Ray, Gamma-Ray, and Neutron Detector Physics X*, vol. 7079, pp. 707911–707915, 2008.

5. H. Strecker, G. Harding, H. Bomsdorf, J. Kanzenbach, and R. Linde, "Detection of explosives in airport baggage using coherent X-ray scatter," in *Substance Detection Systems*, 1993, pp. 399–410.

6. G. Harding, J. Kosanetzky, and U. Neitzel, "X-ray-diffraction computed-tomography," *Med. Phys.*, vol. 14, no. 4, pp. 515–525, 1987.

7. M. S. Westmore, A. Fenster, and I. a. Cunningham, "Tomographic imaging of the angular-dependent coherent-scatter cross section," *Med. Phys.*, vol. 24, no. 1, pp. 3–10, 1997.

8. J.-P. Schlomka, A. Harding, U. van Stevendaal, M. Grass, and G. L. Harding, "Coherent scatter computed tomography: A novel medical imaging technique," in *Medical Imaging*, 2003, pp. 256–265.

9. U. Kleuker, P. Suortti, W. Weyrich, and P. Spanne, "Feasibility study of X-ray diffraction computed tomography for medical imaging," *Phys. Med. Biol.*, vol. 43, no. 10, pp. 2911–2923, 1998.

10. A. Harding, J.-P. Schlomka, and G. L. Harding, "Simulations and experimental feasibility study of fan-beam coherent-scatter CT," *Proc. SPIE 4786, Penetrating Radiation Systems and Applications IV*, vol. 4786, pp. 202–209, 2002.

11. C. W. Cui, S. M. Jorgensen, D. R. Eaker, and E. L. Ritman, "Direct three-dimensional coherently scattered X-ray microtomography," *Med. Phys.*, vol. 37, no. 12, pp. 6317–6322, 2010.

12. A. Castoldi, C. Ozkan, C. Guazzoni, A. Bjeoumikhov, and R. Hartmann, "Experimental qualification of a novel X-ray diffraction imaging setup based on polycapillary X-ray optics," *IEEE Trans. Nucl. Sci.*, vol. 57, no. 5 Part 1, pp. 2564–2570, 2010.

13. D. L. Batchelar and I. a. Cunningham, "Material-specific analysis using coherent-scatter imaging," *Med. Phys.*, vol. 29, pp. 1651–1660, 2002.

14. S. Pang, Z. Zhu, G. Wang, and W. Cong, "Small-angle scatter tomography with a photon-counting detector array," *Phys. Med. Biol.*, vol. 61, no. 10, p. 3734, 2016.

15. G. Harding, "Energy-dispersive X-ray diffraction tomography," *Phys. Med. Biol.*, vol. 35, no. 1, 1990.

16. P. Fratzl, H. F. Jakob, S. Rinnerthaler, P. Roschger, and K. Klaushofer, "Position-resolved small-angle X-ray scattering of complex biological materials," *J. Appl. Crystallogr.*, vol. 30, no. 5 Part 2, pp. 765–769, 1997.

17. C. K. Egan, S. D. M. Jacques, M. Di Michiel, B. Cai, M. W. Zandbergen, P. D. Lee, A. M. Beale, and R. J. Cernik, "Non-invasive imaging of the crystalline structure within a human tooth," *Acta Biomater.*, vol. 9, no. 9, pp. 8337–8345, Sep. 2013.

18. C. M. Agrawal and R. B. Ray, "Biodegradable polymeric scaffolds for musculoskeletal tissue engineering," *J. Biomed. Mater. Res.*, vol. 55, no. 2, pp. 141–150, 2001.

19. G. Harding, "X-ray scatter tomography for explosives detection," *Radiation Physics and Chemistry*, vol. 71, no. 3–4, pp. 869–881, 2004.

20. S. Sidhu, G. Falzon, S. A. Hart, J. G. Fox, R. A. Lewis, and K. K. W. Siu, "Classification of breast tissue using a laboratory system for small-angle X-ray scattering (SAXS)," *Phys. Med. Biol.*, vol. 56, no. 21, pp. 6779–6791, 2011.

21. C. Cui, S. M. Jorgensen, D. R. Eaker, and E. L. Ritman, "Direct three-dimensional coherently scattered X-ray microtomography," *Med. Phys.*, vol. 37, no. 12, pp. 6317–6322, 2010.

22. J. Delfs and J.-P. Schlomka, "Energy-dispersive coherent scatter computed tomography," *Appl. Phys. Lett.*, vol. 88, no. 24, p. 243506, 2006.

23. F. Schaff, M. Bech, P. Zaslansky, C. Jud, M. Liebi, M. Guizar-Sicairos, and F. Pfeiffer, "Six-dimensional real and reciprocal space small-angle X-ray scattering tomography," *Nature*, vol. 527, no. 7578, pp. 353–356, 2015.

24. Z. Zhu, R. A. Ellis, and S. Pang, "Coded cone-beam X-ray diffraction tomography with a low-brilliance tabletop source," *Optica*, vol. 5, no. 6, p. 733, Jun. 2018.

25. D. L. Batchelar, M. T. M. Davidson, W. Dabrowski, and I. A. Cunningham, "Bone-composition imaging using coherent-scatter computed tomography: Assessing bone health beyond bone mineral density," *Med. Phys.*, vol. 33, no. 4, p. 904, 2006.

26. G. L. Harding, "X-ray diffraction computed tomography," *Med. Phys.*, vol. 14, no. 4, p. 515, 1987.

27. "Medipix3 array detector," 2014. [Online]. Available: https://medipix.web.cern.ch/medipix/pages/medipix3.php. [Accessed: 21-Nov-2015].

28. Z. Zhu and S. Pang, "Three-dimensional reciprocal space x-ray coherent scattering tomography of two dimensional object," *Med. Phys.*, vol. 45, no. 4, pp. 1654–1661, 2018.

29. J. M. Feldkamp, M. Kuhlmann, S. V. Roth, A. Timmann, R. Gehrke, I. Shakhverdova, P. Paufler, S. K. Filatov, R. S. Bubnova, and C. G. Schroer, "Recent developments in tomographic small-angle X-ray scattering," *Phys. Status Solidi A-Applications Mater. Sci.*, vol. 206, no. 8, pp. 1723–1726, 2009.

30. C. G. Schroer, M. Kuhlmann, S. V. Roth, R. Gehrke, N. Stribeck, A. Almendarez-Camarillo, and B. Lengeler, "Mapping the local nanostructure inside a specimen by tomographic small-angle X-ray scattering," *Appl. Phys. Lett.*, vol. 88, no. 16, p. 164102, Apr. 2006.

31. L. Weber, M. Langer, S. Tavella, A. Ruggiu, and F. Peyrin, "Quantitative evaluation of regularized phase retrieval algorithms on bone scaffolds seeded with bone cells," *Phys. Med. Biol.*, vol. 61, no. 9, pp. N215–N231, 2016.

32. R. E. Guldberg, C. L. Duvall, A. Peister, M. E. Oest, A. S. P. Lin, A. W. Palmer, and M. E. Levenston, "3D imaging of tissue integration with porous biomaterials," *Biomaterials*, vol. 29, no. 28, pp. 3757–3761, 2008.

33. G. Wang and H. Yu, "The meaning of interior tomography," *Phys. Med. Biol.*, vol. 58, no. 16, p. R161, 2013.
34. F. Noo, R. Clackdoyle, and J. D. Pack, "A two-step Hilbert transform method for 2D image reconstruction," *Phys. Med. Biol.*, vol. 49, no. 17, p. 3903–3923, 2004.
35. Y. Ye, H. Yu, Y. Wei, and G. Wang, "A general local reconstruction approach based on a truncated Hilbert transform," *Int. J. Biomed. Imaging*, vol. 2007, 63634, 2007, www.hindawi.com/journals/ijbi/2007/063634/abs/.
36. X. Jin, A. Katsevich, H. Yu, G. Wang, L. Li, and Z. Chen, "Interior tomography with continuous singular value decomposition," *IEEE Trans. Med. Imaging*, vol. 31, no. 11, pp. 2108–2119, 2012.
37. H. Yu and G. Wang, "Compressed sensing based interior tomography," *Phys. Med. Biol.*, vol. 54, no. 9, pp. 2791–2805, May 2009.
38. J. Yang, H. Yu, M. Jiang, and G. Wang, "High order total variation minimization for interior tomography," *Inverse Probl.*, vol. 26, no. 3, pp. 350131–3501329, 2010.
39. W. H. Richardson, "Bayesian-based iterative method of image restoration," *J. Opt. Soc. Am.*, vol. 62, no. 1, p. 55, Jan. 1972.
40. A. Chambolle, "An algorithm for total variation minimization and applications," *J. Math. Imaging Vis.*, vol. 20, no. 1, pp. 89–97, 2004.
41. Z. Zhu, A. Katsevich, A. J. Kapadia, J. A. Greenberg, and S. Pang, "X-ray diffraction tomography with limited projection information," *Sci. Rep.*, vol. 8, no. 1, p. 522, Dec. 2018.
42. C. Cortes and V. Vapnik, "Support-vector networks," *Mach. Learn.*, vol. 20, no. 3, pp. 273–297, 1995.
43. C. K. Egan, M. D. Wilson, M. C. Veale, P. Seller, S. D. M. Jacques, and R. J. Cernik, "Material specific X-ray imaging using an energy-dispersive pixel detector," *Nucl. Instruments Methods Phys. Res. Sect. B Beam Interact. with Mater. Atoms*, vol. 324, pp. 25–28, 2014.

8

Energy-Resolving Detectors for XDi Airport Security Systems

Smiths Detection Germany GmbH

Dirk Kosciesza

Smiths Detection Germany GmbH

CONTENTS

8.1 Introduction

In this chapter, X-ray diffraction (i.e., XRD) security systems that Morpho Detection[1] (i.e., MD) has developed are introduced. After a historical overview, the functional principle of MD's XRD and its novel X-ray diffraction imaging (i.e., XD*i*) security-system is elucidated, with a focus put on the different detectors developed for these systems, which are based on Germanium, CdZnTe, and CdTe. Their design and functionality are explained, and their performance characteristics are compared to each other. A brief description of the principle of threat-detection in XD*i* precedes the conclusion.

8.2 Physical Principle of X-Ray Diffraction Screening System

Baggage scanners based on X-ray diffraction take advantage of the fact that most explosives are composed of polycrystalline material to a large extent, which enables their investigation by crystallographic methods that typically probe the structure and therefore a very material-specific property of the object under inspection. It is well known from powder diffraction that X-rays irradiating a polycrystalline substance give rise to strong scatter signals, because the periodic structure of the crystallites allows for constructive interference of elastically scattered X-ray photons, provided they are obeying Bragg's law (Warren, 1990),

$$n \cdot \lambda = 2 \cdot d_{hkl} \sin(\theta/2) \tag{8.1}$$

which states, under which scatter angle θ, the interference of scattered photons with wavelength λ, or its n-th harmonic, reflected by one of the many crystal lattice planes with spacing d_{hkl} within the crystallite, can be observed. Equation 8.1 can be rewritten using $E=hc/\lambda$, with h being Planck's constant and c the speed of light, to express the change in the photon's momentum x by the scattering through θ as

$$x = \frac{E}{hc} \sin \frac{\theta}{2} \tag{8.2}$$

with $x=n/(2d_{hkl})$. Hereby h, k, and l are the Miller indices that characterize each lattice plane's orientation in the unit cell. Since explosives typically contain many crystallites oriented in all possible directions, many reflections can be explored at once, if the Laue method of X-ray diffraction is applied (Warren, 1990). Unlike most crystallographic methods, the Laue method employs polychromatic instead of monochromatic X-rays to irradiate the object, in this way exploiting the full available spectral width and flux of photons produced by an X-ray tube. A high flux of photons is crucial for security screening as it aids a faster inspection of the bag, and the broad bandwidth of available photons excites all accessible reflections simultaneously. In other words, it fulfills many Bragg's conditions for different d_{hkl}–λ pairs at once and, in this way, produces several diffraction peaks in the energy spectrum under the same angle. Furthermore, the presence of photons up to energies as high as 180 keV in the source spectrum makes penetration of dense objects possible. As a consequence, a detector should have sufficient absorption up to the highest energies and must be energy resolving to measure the diffraction peaks at different λ, respectively, energy E, under known constant angle θ.

Figure 8.1 shows a simplified diffraction setup. Primary slits produce a highly collimated, but still diverging X-ray beam that interacts with an object (not shown) on the beam-path, producing scatter into all directions. A secondary collimator is positioned under the desired scattering angle θ with a detector behind it that sees all scatter coming from a region in object space, which intersects with the detector's field of view and the primary beam in all three dimensions, forming the scatter voxel.

Hereby, the primary beam divergence and the width of the detector's acceptance angle predominantly define the variation of possible scatter angles $\Delta\theta$ around θ and thereby the achievable momentum transfer resolution but also broadening of the spectral profile due to the detector's energy resolution will become important, if the latter is not sufficient.

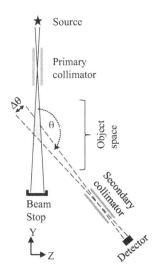

FIGURE 8.1
Typical experimental diffraction setup. The shaded region is the origin of primary scatter photons, i.e., the scatter voxel.

The scattering angle θ must be chosen in such a way that the relevant diffraction peaks of threat and benign substances have significant separation from each other, which is why XRD in security works in the small-angle scattering regime around a few degrees, where a good separation for explosive peaks is achieved. However, at small angles the voxel's height increases proportionally to $1/\theta$, so that for too small angles spatial resolution is lost, potentially resulting in a mixture of threat- and non-threat materials' diffraction signatures within the same voxel.

Therefore, the scattering angle and collimation of both the primary and secondary collimators must be chosen carefully to arrive at an optimal momentum resolution for the explosive-detection problem, keeping aspects of flux and spatial resolution in mind.

It should also be mentioned that non-crystalline substances, even fluids, exhibit characteristic diffraction patterns, owing to the near order of molecules within them, allowing for their identification by XRD as well (Harding & Delfs, 2007).

8.3 History of XRD Security-Scanners and Motivation for Next-Generation XD*i* Systems

Diffraction scanners entered the airport security market at the end of the 1990s, probably inspired by the works of Harding, 1989; Strecker et al., 1993; Luggar et al., 1995. The first system commercially available was the Heimann Diffraction X-ray (HDX), sold by Heimann Systems GmbH, now Smiths Heimann GmbH (Ries, 2001; Smiths-Heimann, 2004). In a multi-level concept (Ries, 2003), threats were identified in a first-level system, and suspicious bags conveyed to the diffraction stage for material-specific alarm resolution. The HDX used a Germanium detector with a collimator in front that could be positioned to point at the suspicious region, which was irradiated by an X-ray pencil-beam to create the scatter signal (Hartick, 2017).

Another system, which had been patented (Harding, 1989) and was developed by Philips, was the XES3000, a system that could inspect 12 height layers of a bag simultaneously by topographic imaging. Philips sold its industrial X-ray business in 1998 as a management buy-out to Yxlon International GmbH, who produced the XES3000 and installed a first system at Duesseldorf Airport (Germany) in 2000. In 2003, the California-based company Invision Technologies, manufacturer of computer tomography systems for baggage inspection, acquired Yxlon, and the XES3000 was improved to become the XRD3500 (see Figure 8.3 top left), which distinguished itself from its predecessor primarily by an imaging pre-scanner and a different mechanical structure. In 2005, General Electric acquired Invision Technologies and integrated it into GE Homeland Protection.

A schematic of the XES3000/XRD3500 diffraction unit is shown in Figure 8.2 to illustrate the functional principle, and the XRD3500's hardware realization is shown in Figure 8.3. A 12 kW rotating-anode X-ray tube on top of the setup generates a high flux of photons that a primary collimator shapes into a hollow cone beam which penetrates the bag, generating scatter at the objects inside, which is then received by a centrosymmetric, multi-slit collimator (see Figure 8.6) guiding the radiation onto a segmented Germanium detector. Each slit images a different layer in height of the object space, so that the 12 different layers are explored at once.

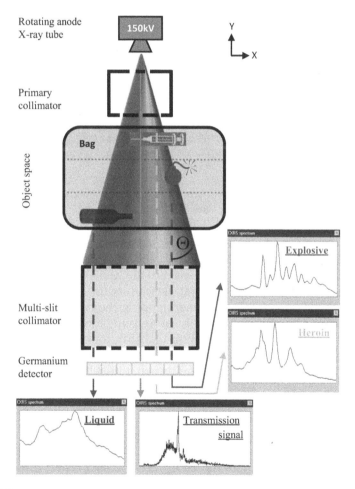

FIGURE 8.2
Schematic side view of XRD3X00 systems. A hollow X-ray cone beam penetrates a bag and produces scatter at objects in different heights of the bag. The scatter is seen by the Ge detector through a collimator under constant angle and produces the diffraction spectra depicted. A transmission beam's spectrum is shown as well.

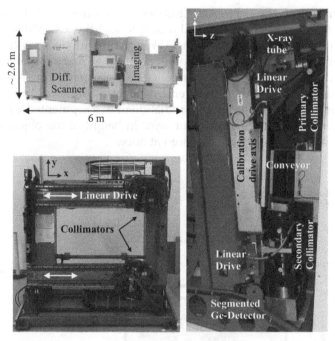

FIGURE 8.3
XRD3500 baggage inspection system (top left), a view onto its diffraction stage along the conveyor direction (bottom left), and a side view of the same (right).

In order to scan the whole bag with this cone beam, the top section, i.e., tube and primary collimator, and the bottom section, i.e., detector and secondary collimator, are connected by a stiff C-arm in the XES3000 that hangs on rails of a strong supportive frame. The bag is scanned in a line-by-line fashion by moving the C-arm along the frame, stepping the bag forward after each line scan. In the XRD3500, the mechanical coupling by the C-arm is removed to reduce the system width from 3.85 m to 2.55 m for better integration into existing baggage-handling systems. This is achieved by letting the top and bottom section be driven on linear motors that are electronically geared to ensure alignment of the X-ray optics (see Figure 8.3 bottom left). While the diffraction-unit moves across the bag, a position-dependent signal repetitively triggers the detector's data acquisition system about every 120 ms to capture the scattered photons passing through the moving secondary collimator before the next trigger arrives. In this way, the scatter is encoded along the x-direction, through the width of the cone in the z-direction and through the topographic imaging in the y-direction to form a voxel in object space. The result is a three-dimensionally resolved scatter distribution of the bag allowing to distinguish different items therein by their material-specific scatter signatures. In addition to the scatter signal, the Germanium detector collects a transmission signal to recognize dense objects in the bag that would attenuate the scatter signal significantly enough to hinder an alarm for a threat object. In this case, the detection algorithm issues a so-called dark alarm. The actual threat identification is accomplished by comparing voxel spectra to spectra from a threat library. Figure 8.2 shows example spectra of substances of interest, which are not limited to explosives but liquids, drugs, and pharmaceuticals can be identified as well.

Both XRD-systems exhibit excellent detection- and false alarm rates, proven by certifications in the US (Morpho Detection, 2015) and EU (ECAC, 2017a and b). Nevertheless, commercial success was limited due to high cost that came with the complex technology, as well as uncompetitive throughput for a standalone system with only 60 bags/hour, whereas Computer Tomography (CT) systems or other imaging systems on the market nowadays inspect up to 1,800 bags/h. That is why the XRD3500 is used in combination with a CT-system in a system-of-systems approach today, resolving the alarms of suspicious bags arriving from upstream CT systems (MorphoDetection, 2012), whereby throughputs of up to 140 bags/hour are reached, depending on the inspection mode selected.

General Electric sold its homeland protection business to the French conglomerate corporation Safran in 2009, which integrated it into their existing security business Morpho as Morpho Detection. Motivated by the intention of the EU to lift the liquid ban for carry-on baggage in 2013 and to deploy checkpoint scanners with capabilities to detect dangerous liquids, Safran enhanced investments into the development of the next generation of X-ray diffraction imaging scanners for inspecting cabin baggage, the so-called XD*i*-CBS, which should be faster, smaller, and cheaper than XRD systems.

For a use at airport security checkpoints, this required a system that could produce a 3D scatter image of a bag at a conveyor belt speed of 15 cm/second and above, hence a scan time of about 4 second per bag considering the typical length of checkpoint trays of 60 cm. Three areas for improvement were addressed to achieve this goal: (i) a new, more efficient system geometry making mechanical movement, except for that of the baggage, unnecessary, (ii) generation of X-rays by a stationary anode X-ray tube with much higher lifetime than a rotating anode tube, and (iii), as a result of the new geometry, a larger detector area to capture more flux.

For (i) different diffraction geometries were investigated (Harding, 2005; Harding et al., 2010) and a decision for the multiple inverse fan beam geometry taken, which demanded, (ii) an extended high-power multi-focus X-ray source, i.e., MFXS, (Harding, 2009), and (iii) an array of room temperature semi-conductor energy-resolving detectors to have an extended detector area at a competitive price (Harding et al., 2009).

Hardware development for the XD*i* started at MD Germany in Hamburg in 2009, and in 2014, a first airport trial with XD*i*-CBS had been conducted at Charleroi airport in Belgium. A further milestone was XD*i*-CBS' EU certification in 2016 (ECAC, 2017). Due to the potential seen in X-ray diffraction imaging, the concept of XD*i*-CBS was extended to a hold-baggage scanner,

XD*i*-HBS, that is essentially an upscale of XD*i*-CBS, and for which a first prototype has been built at Morpho Detection Germany in Hamburg. It has successfully conducted a data collection at the Transportation Security Laboratory near Atlantic City, USA, in spring of 2017.

In 2017, Safran sold Morpho Detection to Smiths Detection, one of the world's leading companies in the security business.

8.4 XD*i* Functional Principle

The goal of XD*i* is to measure a bag's scatter signature at each voxel throughout the bag while it is conveyed through the XD*i* scanner. For the detection of threats, this 3D scatter image must possess sufficient spatial resolution, momentum resolution, and signal strength.

Figure 8.4 shows a side- and front-view of XD*i*'s central part, the diffraction stage, to show how spatial resolution for the height is achieved by topographic

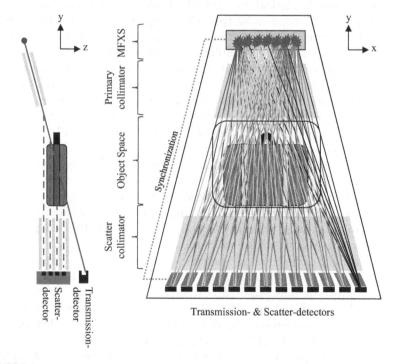

FIGURE 8.4
Schematic view of XD*i*. *Left*: Topographic imaging, one of the primary beamlets (-) generating scatter (- -) through the secondary collimator slits onto the detectors. *Right*: Colored lines show web of primary beamlets for 7 foci, if all foci were switched on at once. Black lines demonstrate how X-rays from different foci converge toward the rightmost detector.

imaging (left), which is applied to many distinct small fan beams, so-called beamlets, whose width defines the resolution along the x-direction (right). Resolution along the z-direction is attained by gating the measurements while the bag is moving forward. Final momentum resolution depends on focus size, detector energy resolution, and the slit widths of the collimation system, which, together with the power and high voltage of the X-ray tube and efficiency of detectors, also determines the achievable signal strength.

Figure 8.5 is a picture of the diffraction stage, a metal frame designed for highest stiffness and least susceptibility to thermal expansion in order to maintain the accuracies of the collimation system and its mounted components. It is tilted, so that the system's overall height will not exceed 2.1 m. On top rests a custom-designed 6 kW MFXS with 16 X-ray focal spots that fire sequentially for 125 µs in order to limit power deposition into each anode button and to allow for heat dissipation through it into the anode heat sink before the electron beam hits it again after a full cycle of 2.1 minutes.

The primary collimation system (Harding et al., 2008) is constructed in such a way that it lets 17 beamlets pass for each active X-ray focus. Each beamlet is directed toward one of the 17 dual energy transmission detectors on the bottom, which also serve as beam stop. This is pictured schematically for seven foci in Figure 8.4 (right), where all beamlets originating at the same X-ray focus have the same color. Equivalently, when viewed out of the

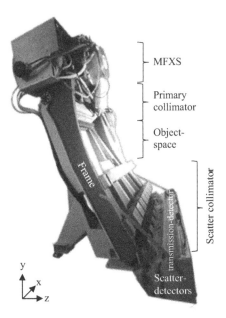

FIGURE 8.5
Photo of the XDi diffraction stage (provided by Dr. T. Theedt). Scatter detectors are inside of the compartment, the transmission detectors are on top of it.

detector perspective, each X-ray focus can be seen by every detector, consequently each detector is located at one of the 17 foci of the primary collimator imaging system to which the rays from the 16 X-ray foci converge, which is shown as black lines for the rightmost detector as example. Hence, the term *multiple inverse fan beam topology* is used to describe XD*i*'s geometry (Harding et al., 2012).

After one complete focus cycle, a web of 16×17 X-ray beamlets has swept through object space in the xy-plane, ensuring full coverage of the tunnel section ($\approx 60 \times 40\,cm^2$), while the bag has moved forward by about 0.3 mm. X-rays interact with the object and are coherently scattered into various directions but only those which pass through the slits of the secondary collimator under a small constant scattering angle will reach the scatter detectors (see Figure 8.4 left). The secondary collimator resides underneath the conveyor and consists of three layers of 17 stacks of parallel thin metal plates, each forming 22 channels between them. Consequently, each detector is divided into 22 segments along the z-direction, each imaging a voxel in object space. Hence, whenever a focus fires, 17×22 different scatter spectra are captured, so that after a full focus sweep 5984 spectra have been acquired making up one slice of the 3D image, of which there would be ~1800 in total for a bag of 60 cm length corresponding to 11×10^6 spectra. If the segments are subdivided further, e.g., when a pixelated detector is used, the number of spectra will be multiplied by the number of pixels per segment, i.e., currently eight for XD*i*-CBS.

In order to allocate each spectrum correctly in time and space, the X-ray source sends a signal to a trigger logic each time a new focus fires, which then generates a tag including focus number and belt position to pass it on to each detector. This tag is then attached to each photon measured during the dwell time of the focus, together with a respective detector and channel number. In this way, each photon's origin from within the bag is determined with a spatial uncertainty that corresponds to the size of a sub-voxel, which is formed by the intersection of the beamlet with the field of view of the detector channel. The distance the bag moves at this location during the dwell time of 125 µs is negligible with just 18.8 µm, but the bag also moves forward during the firing of the other foci so that a dead space of 335 µm follows the sub-voxel associated with a specific beamlet, detector channel, and time stamp. However, this dead space may be crossed by another beamlet–detector–channel combination, so that effectively there is no dead space of mentionable size following a beamlet.

Each detector aggregates the tagged photons into data packets and sends them to the data-processing computer via UDP then. Before any spectrum is built from the photons, they are first projected back into a chosen voxel space, whereby each of these voxels unites those sub-voxels that fall into their location.

The respective photons must be transferred into momentum space first by making use of the scattering-angle associated with each sub-voxel, before they can be added together to form the final voxel spectrum that is available to the subsequent detection algorithms.

8.5 Detector Technology in X-Ray Diffraction Security Screening

Seven requirements that a detector for an XRD-scanner has to fulfill have been identified: (i) the spatial resolution of the detector must fit to slit widths and shapes of a multi-slit collimator, (ii) detection efficiency must be high for the energy-range chosen, (iii) energy resolution should be comparable or less than the angular resolution of the diffraction system, (iv) the count rate capability of the detector must be able to handle the occurring fluxes, (v) it must be capable of being integrated into an airport environment, (vi) must have high reliability and preferably low maintenance, and (vii) of course, cost matters.

In the following, the detectors used in XRD3X00 systems are described first, then those developed for XDi systems.

8.5.1 Germanium Detectors of XRD3X00 Hold-Baggage Screening Systems

When the XRD3000 was designed at the end of the 1990s, it seemed that high-purity Germanium detectors were the best option: lithographic segmentation to match the collimator pattern was possible, thickness could be chosen to match the desired absorption, excellent charge collection efficiency guaranteed high peak efficiency, and energy resolution for a segmented detector was still very good around 1 keV using discrete data-acquisition electronics that could handle count rates of 15 kCts/s per channel. However, to utilize these favorable properties of Germanium detectors, cryogenic cooling of the crystal in a high vacuum chamber is necessary. Typically, liquid nitrogen was used for cooling Germanium detectors, but there is neither an infrastructure for liquid nitrogen at an airport nor the willingness to deal with associated maintenance, downtime, safety issues, and cost. Fortunately, mechanical coolers based on the Joule–Thomson (JT) principle were available, e.g., the Cryotiger by APD (then Polycold, now Brooks Automation). Its cooling power at –175°C was around 10 W, but drops off with time when the fine capillaries in the cold-head clog with contaminants from the gas stream, e.g., like precipitated oil-dust. To prevent this, maintenance procedures that required warming up the detector every 6–12 months were necessary resulting in downtimes of more than 24 h because of thermal cycles and associated vacuum evacuation of the detector's cryostat. With the introduction of the PCC as successor of the Cryotiger in 2006, its remarkably improved performance reduced these problems considerably (Brooks Automation, 2013).

Therefore, requirements (i)–(iv) were well fulfilled by a Ge detector system, item (v) acceptably and items (vi)–(vii) could be accepted the way they were although it was hard to accept a price in the low six-digit range for the bill of materials.

The special detector that is still used for XRD3X00 systems was designed and manufactured by Eurysis Mesures (Canberra thereafter, now Mirion Technologies). It is a planar, high purity, low-energy Germanium detector with 2×12 circular symmetric segments (i.e., EGPS) to capture scatter radiation and four channels for transmission. A corresponding layout of the bottom diaphragm in the secondary collimator is shown in Figure 8.6. Another design challenge was to make the detector withstand the scanning motion with accelerations of up to 1 g.

The first generation of detectors had an integrated cold-finger that received its cooling gas through 5 m long tubes that connect it to the compressor of the Cryotiger system, respectively PCC, generating a pressure of 19 bar within the refrigeration system. The special made corrugated metallic gas tubes have to withstand constant bending as they follow the motion of the scanner. At the bottom of the detector sits a valve that allows for attaching a vacuum pump to evacuate the detector chamber. Inside is an adsorption pump for maintaining the vacuum, JFETs for signal pre-amplification that also generate a heat load in the cryostat, and, underneath the Beryllium entrance window on top, the Ge crystal with a thickness of 10 mm. The cryostat is surrounded by further electronics for the pre-amplification of signals, which are fed out through two massive connectors aside the cold-finger into 9 m long, well-shielded cables to connect with a distant 28-channel data acquisition system. Throughout the years, commercial analog and digital systems from target System Electronic Gmbh (Stein, 2017), Silena (not existing anymore), and, for the latest XRD-systems, CAEN (CAEN S.p.A, 2015) have been in use, whereas changes of vendor were driven by price advantages or product discontinuation.

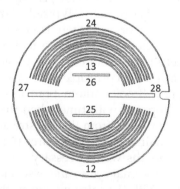

FIGURE 8.6
Layout of the diaphragm in front of the Ge detector. Slits 1–24 collimate scatter radiation, slits 25–28 receive attenuated transmission beams.

A big improvement for the detector was its redesign to replace the Joule–Thomson cooling with pulse tube (i.e., PT) cooling technology, which started to be integrated into Germanium detectors at Canberra around 2004, when it became available at the cooling power needed (~5 W) at –175°C. A PT cooler is a hermetically sealed refrigeration system consisting of a compressor to pump the cooling gas, e.g., Helium, and a connected cold head, the pulse tube, which is free from moving parts to eliminate wear and vibrations that could otherwise deteriorate the detector's energy resolution. The compressors have a double-piston design for cancellation of associated vibrations and their pistons hang on flex bearings resulting in a maintenance-free system with MTTFs of 90,000 h (Benshop et al., 2002; Thales Cryogenics, 2017). The first PT-cooled XRD detector was delivered in 2005 and is depicted in Figure 8.7 together with its characteristics in Table 8.1. It achieved an improvement in average energy resolution from 0.85% to 0.74% at the 122 keV[57]Co-line over the JT-cooled detector, but its main advantage is its reliability and the cooling system's compactness

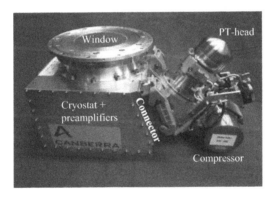

FIGURE 8.7
EGPS Ge detector with pulse tube cooling.

TABLE 8.1

Key Features of EGPS Ge Detector

Key Features	Characteristic
Product	EGPS 90 N 2 × 12 +4
Energy range	1–160 keV
Crystal size	Ø 100 mm × 10 mm
Number of segments	24 (1224 mm^2) + 4
Inputs	±12 V, 1000 V
Primary/secondary cooling	Pulse-tube (–175°C)/water (20°C)
MTTF pulse tube	90,000 h
Readout	External discrete DAQ
Acceleration	≤1 g sinusoidal
Enclosure dimensions	420 × 240 × 293 mm^3

that lets it fit onto the detector itself without the need for an external compressor. The achieved resolution of the two different detector-types is histogrammed for 24 JT- and 12 PT-detectors in Figure 8.8. Three detectors can be identified as outliers with regard to their performance; however, they are still well below the resolution requirement of 1.8 keV at 122 keV, 1.5% respectively. Figure 8.9 shows the energy resolution of each of the 28 segments averaged over 36 detectors with error bars indicating the standard deviation, which is ≈0.1keV. Energy resolution increases from inner to outer segments because of increasing capacitance with segment size (see Figure 8.6).

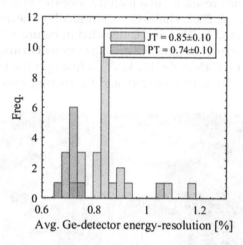

FIGURE 8.8
Distribution of average energy resolution for Joule–Thomson and pulse tube cooled Ge detectors.

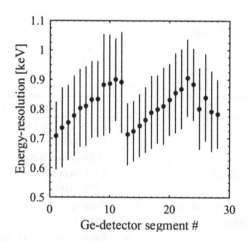

FIGURE 8.9
Mean energy resolution per segment for 36 detectors with standard deviation.

8.5.2 Energy-Resolving Room-Temperature Semi-Conductors for XD*i* Systems

As outlined above, Germanium detectors seem to be the first choice for energy-dispersive XRD applications based on their performance. However, when price starts to matter, one is willing to look for alternatives, especially, when many detectors are needed, as in the case of the XD*i* design. The price of the current Ge detector is already prohibitive for the bill of materials of XD*i*, and this would probably also be the case for a system with 17 Ge crystals, associated cryostat(s), cooling technology, and the data acquisition system.

Since the individual detector areas are evenly distributed over a width of 1.36 m in XD*i*-CBS' detector compartment, they would either have to rest in a correspondingly big vacuum tank, making any repairs in the field nearly impossible or in smaller tanks that could be exchanged in a service case. Nevertheless, cooling down an exchanged detector can easily take half-a-day, adding to the downtime of a defective system, an unfortunate circumstance for the operation of an airport checkpoint. Furthermore, the number of necessary distinct detector channels in XD*i* amounts to 374, requiring at least just as many data acquisition channels that would have to be realized with ASICs for cost reasons as discrete electronics are expensive. Principally, ASICs for Germanium detectors are available and combinations have been realized, (Shih et al., 2012; Rumaiz et al., 2014; IDEAS, 2017), but do not seem to be a frequent solution.

Hence, energy-resolving room-temperature semi-conductor detectors (i.e., RTSD) were investigated as an alternative, with Cadmium Telluride (CdTe) and Cadmium Zinc Telluride (CZT) being the technically most mature candidates judging by their achieved energy resolution.

CdTe and CZT are materials with excellent stopping power in the energy range of interest due to their high average atomic number ($\approx$50) and density ($\approx$5.8 g/cm^3). Figure 8.10 shows the absorption of CZT, CdTe, and Germanium for material thicknesses as used in XRD/XD*i* detectors (calculated with XOP (Sanchez del Rio & Dejus, 2004)). Above 90 keV absorption starts to drop, with 5 mm thick CZT and 4 mm CdTe maintaining a little higher absorption than 10 mm Ge, until absorption reaches 68% at the end of the range of interest at 180 keV, and around 62% for Ge and CdTe. Apparently, material thickness of RTSDs was chosen to be comparable in absorption with 10 mm of Ge.

RTSDs have a bandgap of 1.5 eV and a pair-creation energy of $\approx$4.5 eV, which are about 2.2 resp. 1.5 times higher than for Germanium, allowing for operation as a radiation detector at room temperature, because thermal excitation of electrons into the conduction band as described by Boltzmann's equation is much less at room temperature than for Ge with its pair creation energy of 2.95 eV. This comes at the price of a reduced signal to noise ratio, as less electron–hole pairs are created for the same energy deposited in the crystal, e.g., 20,180 vs. 13,230 electrons at 59.5 keV. Furthermore, charge collection properties are inferior, reflected in the mobility lifetime products $\mu\tau$ for electrons

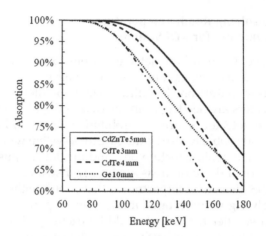

FIGURE 8.10
Absorption of the detector crystals used for XRD/XD*i* detectors.

and holes that are 3–4 orders of magnitude lower than those of Germanium, resulting in incomplete charge collection. In order to compensate somewhat for these reduced $\mu\tau$ values, a high-depletion voltage helps, but this also raises the dark current with a detrimental effect on energy resolution. This problem increases, when thick devices for efficient absorption of higher energies need to be used. One solution is to apply Schottky barrier contacts to the CdTe crystal in order to suppress the dark current when applying high voltage (Takahashi et al., 1998), roughly 800 V per mm crystal thickness are used. Another solution to this problem was to crystallize CdZnTe by replacing 10% of the original Cadmium content by Zinc, raising resistivity by an order of magnitude to over 10^{10} Ωcm (Awadalla et al., 2010), allowing for thicker material at less depletion voltage. But this crystal structure is more complex and therefore even more difficult to grow, which reflects itself in imperfect crystals exhibiting twinning, inclusions, dislocations, and precipitations (Szeles & Driver, 1998; Bolotnikov et al., 2008), resulting in slightly worse charge transport properties when compared to CdTe, which can suffer from these crystal imperfections as well. Hence, it is an art to grow such crystals reproducibly at spectroscopic quality, in high volume and at an attractive price, which is why only a few pioneering companies are producing these materials at industrial scale, the most prominent may be Acrorad (Siemens became majority shareholder in 2011), Redlen, eV-products (acquired by Kromek in 2013), and Imarad Imaging System (acquired by Orbotech in 2005). Nowadays, crystals are grown in ingots up to 4″ diameter, making the material available at industrial scale (Shiraki et al., 2007; Funaki et al., 2007; Chen et al., 2007). Many applications of these materials are in the astronomical, medical, and security field, and many innovative methods have been developed to circumvent or compensate existing problems and create application-specific solutions that achieve good results with regard to energy resolution, for example (Takahashi et al., 2002; Gostilo et al., 2002; Kuvvetli & Budtz-Jorgensen, 2005; Montemont et al., 2007).

Next to the capability to achieve good energy resolution and efficiency for X-rays without cryogenic cooling, technologies for bonding CZT and CdTe crystals to substrates or directly to ASICs have evolved, which allow for having many individual detector segments at low cost making X-ray imaging with small pixel sizes down to 55 μm possible (Procz et al., 2013). The XDi concept does not need such small pixel sizes, besides, discrete electronics for at least 390 electronics channels would already be beyond the financial scope of such a system, hence availability of low-noise ASIC-technology is crucial to the success of RTSD in XDi.

An important question was, whether RTSD's energy resolution would be sufficient for XDi. In answering this question, it needs to be considered that the final width of diffraction peaks in energy space results from the angular broadening caused by the allowed divergence of rays in the collimation system that hit the object and the acceptance angle of the secondary collimator and the detector's resolution, hence (Cozzini et al., 2010):

$$\Delta E_{\mathrm{Peak}} = \sqrt{\Delta E_{\mathrm{geometry}}^{2} + \Delta E_{\mathrm{det}}^{2}} \tag{8.3}$$

Figure 8.11 illustrates the influence of different detector resolutions (1%–5%) on final diffraction peak resolution for different geometrical resolutions of 4%–9%. It can be seen, how the influence of worsening detector resolution reduces as geometrical resolution increases from 4% to 9%. For example, the Ge detector's resolution at 59.5 keV is 2.2% and broadens a diffraction peak

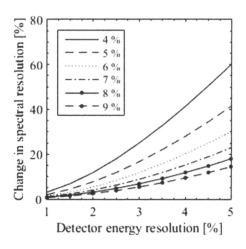

FIGURE 8.11
Change in spectral resolution caused by different detector resolutions from 1% to 5% for given geometrical resolutions of 4%–9%.

of 4% by only 14%, i.e., to 4.6%. Replacing the detector by one with 3% or 4% resolution broadens it by 25 resp. 41% to already 5 resp. 5.6% total peak resolution. For the development of XD*i*'s detectors, a target energy resolution of 3% and an upper limit of 4.5% at 60 keV was set and considered for the design of the collimation system.

8.6 Two Approaches to RTSD for XD*i*

Motivated by impressive energy resolutions achieved with CdTe detectors down to 2.5% at 60 keV (Takahashi et al., 1998), MD investigated the possibility to use CdTe for XD*i*. Since a high voltage of around 800 V per mm crystal thickness needs to be applied across the diode for good charge collection, only sensor thicknesses up to 2 mm are feasible, too low of a thickness for efficient absorption of X-rays up to 180 keV.

However, if the sensor is rotated by 90°, the absorption length will increase to the width of the sensor while it can still be biased along its smallest dimension, profiting from the small drift path for the created charges as the photons hit the crystal *edge on* now (see Figure 8.12, left). This is an elegant approach, but problems arise when detectors covering a larger area are needed, because it requires stacking. Then the tilted crystals need to maintain a spacing between them to prevent potential arcing of high voltage and to provide space for contacts and electronics. Consequently, there will be a

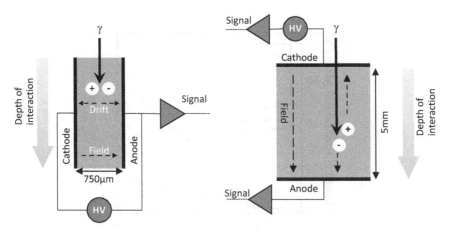

FIGURE 8.12
Left: Schematic of edge-on geometry, field is perpendicular to incoming photon direction, hence, the drift path of charges is always less than 750 μm, independent of interaction point. *Right*: Face-on geometry, field is parallel to incoming photon direction, hence, length of drift path can vary from 0 to 5 mm depending on interaction point.

dead space in between individual crystals, which is no obstacle for XD*i* as the dead space is positioned underneath the collimator septa that separate the sensing areas of the detector.

In 2009, a feasibility study to assess the achievable energy resolution for an edge-on detector in collaboration with the American detector development house DxRay started and entered into a development project in 2010 with the intention to develop a CdTe edge-on array detector composed of 22 individual crystals.

A second detector development was started together with the French innovation and technology transfer institute CEA-Leti, also in 2010. The technology proposed is built upon a 5 mm thick CZT crystal in *face on* geometry, i.e., with the electric field parallel to the incoming photon direction (see Figure 8.12, right), so that photon-generated charges, depending on their interaction point, may have to travel up to 5 mm through the crystal. In order to alleviate the deterioration of energy resolution due to incomplete charge collection that appears in CZT, employing the so-called *small pixel effect* (He, 2001), charge-sharing and depth-of-interaction corrections (i.e., DOI) (Verger et al., 2004) are applied. The first separates the crystal lithographically into smaller pixels to arrive at a favorable ratio between pixel size and sensor thickness for the charge-induction process, the second analyses the charge cloud for charges that have migrated to a neighboring pixel and adds them back to the pixel that fired and the last correction compensates for interaction-point dependent charge-induction loss, which requires to measure both anode- and cathode-signal of a photon event.

The requirements for the two different detector developments were set as equal as possible to facilitate system integration into XD*i* and interchangeability of detectors. Mechanical dimensions, alignment pins, and possibilities for fixtures of the detector and a cooling plate to it were mandatory to be the same, as well as the data structure and the data transmission protocol. Communication with regard to control and monitoring of the detector had to be realized via TCP/IP, data transmission via UDP protocol. Attention was paid to potential electromagnetic interference from the airport environment, which is why the robust RS-485 standard was chosen for trigger transmission and high voltage had to be generated internally to minimize the length between high-voltage supply and the crystal. Hence, the detectors just need their low-voltage supply, the trigger, and one Ethernet connection for control and data transmission to operate. The two detectors are presented in more detail in the following.

8.6.1 The CdTe Edge-On Array Detector (EdgAr)

The CdTe edge-on array detector for XD*i*-CBS (EdgAr) developed by DxRay (Iwanczyk et al., 2011; Iwanczyk et al., 2014) has a detector head equipped with 22 CdTe crystals, which was however adapted during the project to accommodate 26 crystals in order to suit XD*i*-HBS with its higher tunnel height as well. This detector version is named EdgAr+.

The crystals have a thickness of 750 µm over which the high voltage is applied, the absorption depth is 3 mm for EdgAr and 4 mm for EdgAr+, and their width is 22 mm, along which the sensor is segmented into 15 pixels to have less capacitance per pixel for achieving better energy resolution (see Figure 8.13 top left). Principally, spatial resolution along this direction is not required, which is why these 15 spectra are added together in the data processing to form a segment spectrum.

Each crystal is mounted to its own carrier board, the "array card," equipped with connectors and the low-noise ASIC XrM15, which was custom designed for this application in CMOS 0.35 µm technology by DxRay.

Its input stage consists of preamplifier, shaper, peak detector, and a track and hold circuit that is followed by a comparator to generate trigger signals and a DAC is available for setting the comparator threshold. The chip also possesses a multiplexer and a sequencer for organizing the read out.

The array cards are stacked onto each other to form a sensitive area of 22 mm × 27 mm, respectively, 30 mm for HBS, whereby two crystals are separated by a dead space of 500 µm at a pitch of 1.25 mm (see Figure 8.13, bottom left), for which a precision of ±50 µm must be maintained to ensure alignment with the collimator. The position of the stack relative to the collimator is aligned by pins in the stack holder that sticks through the detector enclosure and are inserted into precise receiving holes in the collimator.

Each array card has a flat ribbon cable leading to one of the two receiver boards left and right of the array card stack. Each receiver board has 14 connectors (one as spare) for the ribbon cables, three ADCs, and one FPGA that organizes data readout, digitization, and buffering. The receiver boards send acquired data to the motherboard, on which a system-on-a-chip board runs a UNIX-operating system to handle calibration of data, its packaging and transmission, as well as communication of control tasks and monitoring. A power board generates all necessary voltages from the 5 and 2.8 V supply voltages, including the bias voltage. In order to prevent the problem of

FIGURE 8.13

Top left: Photo of segmented CdTe crystal (provided by Acrorad). *Bottom left*: Photo of stack of array cards with crystals (provided by DxRay). *Right*: CZT detector head board manufactured by Redlen.

polarization, which can be observed for CdTe detectors with Schottky barrier contacts, a bias-reset technique with a 0.5% duty cycle is applied (Seino & Takahashi, 2007).

8.6.2 The CZT Imaging Diffraction Detector (IMADIF)

The CZT imaging diffraction detector (IMADIF) has been developed by CEA-Leti according to XD*i*'s requirements and consists of a detector head board, underlying ASIC boards, an ADC board, a trigger board, a motherboard, and a power board that generates all necessary voltages including the bias voltage (Montémont et al., 2013).

For reasons of cost and yield, the detector head board consists of three, well-aligned 5 mm thick CZT tiles of spectroscopic quality, arranged to form one sensitive area of 22×30 mm^2. The anode side is segmented into 24×8 pixels that are directly bonded to the underlying PCB, but only the inner 22 lines resp. 176 pixels are used. The cathode side is segmented into 3×4 large cathodes that are contacted by a flex board covering the cathode (Figure 8.13 right). The detector head board is a joint development between Leti, Morpho Detection, and Redlen Technologies, who is also the manufacturer of it.

In this development, Leti used IDeF-X HD (Gevin et al., 2012) as anode ASIC, which had been developed by CEA-Saclay for space applications, and ALIGASPECT for the cathode side, a 16-channel ASIC developed by Leti for X-ray spectrometry (Rostaing, 2017). All ASIC boards are connected to one underlying ADC board, which interfaces with the motherboard that controls the ASICs, reads out the data, and performs charge sharing and depth-of-interaction corrections. For this task, it possesses an FPGA with a NIOS embedded processor, on which a simplified UNIX operating system runs. The trigger board is attached to the motherboard as well.

Temperature information is important to verify that the detector is working within its specified operating range of 20°–25°, which is why IMADIF provides readout of ASIC temperatures and a temperature sensor close to the crystal. Furthermore, the values of all generated voltages of the power board are available for troubleshooting.

A distinctive feature of the detector that aids manufacturing and quality control is its automatic self-test, after any important manufacturing step. There is a verification test for power board voltages, for pulser tests on ADC and ASIC boards, adjustment of pixel thresholds, dark field, and bi-parameter calibration, as well as the final spectral test. The bi-parameter calibration requires illumination of the detector with a preferably strong ^{57}Co-source to generate the correction factors for DOI and charge-sharing onboard of the detector.

An overview of the key features of both detectors is summarized in Table 8.2.

TABLE 8.2

Key Detector Features of RIMADIF and EdgAr+

Key Features	RIMADIF	Edgar+
Crystal	CdZnTe	CdTe
Energy range	5–200 keV	10–180 keV
Crystal size	$22 \times 28 \times 5\,mm^3$	$22.5 \times 0.75 \times 4\,mm^3$
# of pixels	192 (24 × 8)	390 (26 × 15)
Pixel size	$3.4\,mm^2$	$1.1\,mm^2$
Inputs	12 V, ethernet, trigger	5 V, 2.8 V, ethernet, trigger
Power consumption	4.5 W	6.3 W
Communication	TCP/IP, UDP	TCP/IP, UDP
Readout ASICs	IDeF-X/ALIGASPECT	XrM15c
Enclosure dimensions	$185 \times 78 \times 75\,mm^3$	$185 \times 79 \times 83$
Specialty	Charge-sharing & depth of interaction correction	Small drift path geometry
Service port	HDMI	3.5 mm cinch
Units produced	80	21

8.7 Characterization of Detectors

In total, 120 CZT detectors were manufactured for XD*i*, 40 by Leti (IMADIF) and 80 by Redlen (RIMADIF), who modified the design slightly for manufacturability to reduce cost and eventually improve performance. So far, only RIMADIF detectors have been installed into XD*i* systems and have proven to deliver the performance necessary to pass the C1 EU certification for cabin baggage scanners.

Of the CdTe detectors, there were five prototypes built for XD*i*-CBS, and in a first small-scale production, 21 EdgAr+ detectors were built by DxRay together with industrial partners.

In the following, the commercially built detectors are compared, together with measurements of one JT-cooled EGPS as golden standard. Spectral quality is investigated based on Americium and Cobalt spectra that have been collected in the same experimental setup for all RTSDs. Temperature stability tests have been performed as well.

Before the results are described, the experimental setup for testing the XD*i* detectors is explained below, and it follows a discussion of the spectral parameters calculated.

8.7.1 Experimental Conditions of Source Measurements

The performance of the detectors is determined by measuring the emission lines of [241]Am at 59.5 keV and [57]Co at 122.1 keV. The sources themselves are in a capsule with a 265 μm steel window that filters low-energy emission, and each capsule resides in a massive collimator with a shutter that opens when it is inserted into the source holder, so that radiation is released down a lead-walled tube to irradiate the detector 358 mm below. The initial source activities were 74 MBq and 370 MBq, create a photon flux on the detector surface of 7.8 γ/s/mm² for ^{241}Am, and 196 to 32 γ/s/mm² for ^{57}Co considering its decay with a half-life of $\tau_{1/2}$=271.8 days during the time span of 24 months, in which the measurements took place. The RTSD rests on a plate cooled by water, whose temperature is regulated by a Peltier cooler. Waste heat is removed from the electronics in this way, ensuring a stable crystal temperature. A temperature sensor is attached to the side wall of the enclosure as feedback for the regulator. It is possible to adjust the regulator's filter parameters in such a way that the measured enclosure temperature at this point is very stable with a variation of σ< 0.1°C around its target temperature, which is 21°C for RIMADIF and 18.5°C for EdgAr+, as the latter detector has a higher power consumption and heat generation.

8.7.2 Spectral Parameters

In order to compare the performance of detectors among each other, their spectra are investigated with regard to the following characteristics: (1) energy resolution (*ER*), (2) peak-to-total-ratio (*PTTotR*), and (3) intrinsic peak-efficiency (*IPE*).

A good *energy resolution* is essential for an XD*i* security application, because it is necessary to find peaks above background or resolve close peak neighbors to discriminate between a peak from a threat or benign material. However, energy resolution should not be achieved at the cost of efficiency, if possible, since photons are scarce in a highly collimated diffraction setup like XD*i*. There are in fact some methods to improve energy resolution in CdTe and CZT detectors, which however have a loss in efficiency as consequence, e.g., using very thin detectors (Takahashi et al., 2001), or signal discrimination methods (Niemelä et al., 1996). Since the number of measured photons in XD*i* translates into the belt speed one can achieve with the baggage scanner at given detection and false alarm rate, discrimination methods to achieve better energy resolution are discouraged.

Peaks in the spectra of RTSDs unfortunately do not have Gaussian peaks like Ge detectors but lose photons from the peak into a tail on the low-energy side of the peak (see Figures 8.14 and 8.15), whereby information is lost as these signals will disappear in the continuum of an X-ray tube spectrum. This tailing is dependent on many factors, for example, type and quality of crystals, electrode-configuration, geometry (edge-on, face-on, pixel size, thickness), energy of the entering photon, and corrections possibly applied. A measure that characterizes this tailing in a sense is the *peak-to-total ratio* (*PTTotR*) that relates the number of photons in the peak region to all counts in the spectrum, which consist of the peak photons themselves, photons with incomplete charge collection that could not be corrected well enough

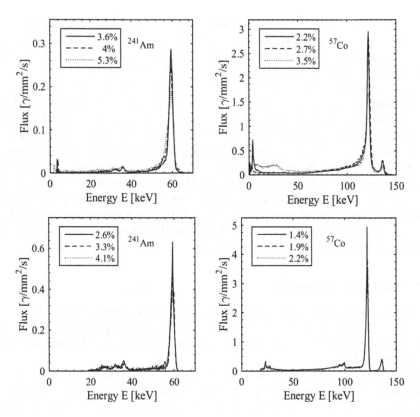

FIGURE 8.14
[241]Am- and [57]Co-spectra of the best RIMADIF- (top) and EdgAr+ detector (bottom) in the deliveries. The best, medium, and worst pixel spectrum of the measurement are shown.

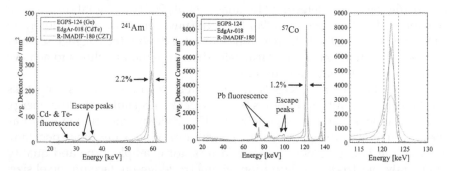

FIGURE 8.15
Spectral comparison of selected EGPS, EdgAr+, and RIMADIF detectors. The average detector spectrum, i.e., the spectrum integrated over 24, 390, 176 channels, for measurements of 600 s duration are shown, normalized to 1 mm² detector area with a bin-width of 0.2 keV/bin. *Left*: [241]Am. *Center and right*: [57]Co. Dashed lines indicate the peak region used for calculation of spectral parameters.

to end up in the peak region, signals from Cd and Te characteristic escapes, scattered photons from the surroundings, natural background and characteristic photons from shielding material, or other emission lines of the source. In the calculations performed here, a spectral range from 20–132 keV is chosen to exclude most of Cobalt's 136 keV photons. Hence, the *PTTotR* determines that portion of photons that remains useful after the detection process and is therefore a measure for the quality of a detector in a given setup.

The *intrinsic peak efficiency* relates the counts in the photo peak to all photons that have actually impinged on the detector area. In order to determine *IPE* and *PTTotR*, the peak range considered to carry valuable information for XDi must be defined. Taking the spectra of the EGPS Ge detector as baseline, and assuming that all photons of a Gaussian peak carry information and are thus valuable, and because the Cobalt peak of the detector has a full width at half maximum (*FWHM*) of 1.2% and is nearly Gaussian, a range of ±2.6σ or 99% of the photons around the photo peak is selected as peak range, corresponding to a window of 2.63% resp. 3.2 keV around the peak. In analogy, a peak range of 4.9% resp. 2.91 keV around the Am peak is chosen.

The calculation of these three spectral parameters, which are investigated for the first production units further below is specified in the following (derived from Knoll, 1989),

1. Energy resolution (*ER*) [%] $\quad:\quad \dfrac{\text{FWHM}}{E^{\text{Peak}}} \times 100$

2. Peak-to-total-ratio $\left(PTTotR\right)$ $\quad:\quad \dfrac{\sum\limits_{i=E_{\text{low}}}^{i=E_{\text{hi}}} y_i}{\sum\limits_{i=E_{\text{Thresh}}}^{i=E_{\text{max}}} y_i}$

3. Intrinsic peak efficiency $\left(IPE\right)$ $\left[\%\right]$ $\quad:\quad \dfrac{\sum\limits_{i=E_{\text{low}}}^{i=E_{\text{hi}}} y_i}{\Delta t \times A_{\text{Pixel}} \times I_{\text{DetSurf}}} \times 100$

whereby y_i is the number of counts in the *i*-th energy bin in the spectrum of 1024 bins with a bin width of 0.2 keV, E^{Peak} the energy bin with most counts, and E_{max}=132 keV the end of the spectral range considered. The threshold E_{Thresh} serves to eliminate any noise from the calculations and is pragmatically set to 20 keV, as diffraction signals there under are not of interest. The peak range is defined from E_{low} to E_{hi} according to the considerations mentioned above, i.e., 58.09–61.00 keV and 120.46–123.66 keV. The pixel area is denoted with A_{Pixel}, the duration of measurement as Δt, and the number of photons per mm^2 and second impinging on the detector surface as I_{DetSurf}.

8.7.3 Spectra of CZT and CdTe Detectors

MD conducted the spectral measurements on the detectors as incoming goods inspection, and Figure 8.14 shows corresponding spectra for the best RIMADIF- (top) and EdgAr+ detectors (bottom), whereby the best, an average and the worst spectrum per detector are plotted with their corresponding energy resolution given in the legend.

RIMADIF's best Americium spectrum reaches an energy resolution of 3.6%, the worst pixel has 5.3% (see Figure 8.14 top left). The spectra exhibit tailing down to about 45 keV, and below 40 keV escape peaks of Cd and Te are visible. The discriminator threshold is as low as ≈4 keV, confirming the low noise of the electronics. Looking at the Am spectra of EdgAr+ (see Figure 8.14 bottom left), an energy resolution about 0.8% better (absolute) than for RIMADIF is observed, and energy resolution varies from 2.6% to 4.1% among the pixels. The peak shape is more similar to a triangular peak than a Gaussian one and has a sharper peak. In addition, the spectra exhibit less tailing below the peak down to 40 keV, where escape peaks start to appear as well.

RIMADIFs Cobalt spectra (top right) exhibit long tailing below the 122.1 keV peak down to about 50 keV, from where on the spectrum is rather flat. However, depending on the pixel's quality, the background may possess a higher level. Quite apparent is the high-energy tailing above the main peak that even stretches out to the 136.5 keV peak. This is a result of overestimated DOI correction. Energy resolution for the best detector varies from 2.2% to 3.5% among the pixels.

For EdgAr+, the Cobalt spectra stick out with steep edges (bottom right), exhibiting no high-energy tailing and a drop to a plateau of about 3% peak height on the low energy side, which extends to ≈100 keV, where the escape signals start to appear at 99 and 97 keV, which are hardly observed in the RIMADIF spectra. They are apparent in the edge-on geometry using thin crystals, because an escape photon is never farther away from the crystal limits than 375 µm and has therefore a higher chance of escaping the crystal than in the case of the 5 mm thick CZT crystal. Upon their escape, the characteristic Cd, respectively, Te $K_{\alpha/\beta}$-photons may be absorbed by a neighboring crystal and appear in its spectrum at their characteristic energies between 23 and 31 keV (Iwanczyk et al., 1979; Bearden 1967). This can be seen for EdgAr+ as some small peaks right above the threshold. The associated peak efficiency loss due to escapes is about 9.5% for EdgAr+ and 5% for RIMADIF.

For direct comparison of the different technologies, the mean spectra of an EGPS, an EdgAr+, and a RIMADIF detector, with their 24, 390, and 176 spectra averaged, are shown in Figure 8.15. The spectra are normalized to 1 mm² detector area and 1 s exposure, and Cobalt measurements are scaled to the initial Cobalt activity.

Due to the large segments of the Ge detector, up to 190 mm², and the resulting high count rate, the Co-source had to be attenuated, first by a 0.5 mm and then in a second measurement by a 1.0 mm thick lead piece to reduce the flux of 122 keV photons by a factor of 6.8 resp. 46.5. Otherwise, considerable dead time effects occur in the data acquisition path. Consequently, the Ge detector measurements are scaled up to compensate for the absorption of lead at 122 keV.

Figure 8.15 (left) shows the Americium spectra with the peak of Ge clearly dominating, being nearly twice as high as the one from CdTe, whose high-energy peak edge is almost as sharp as Germanium's, but the tailing on the low-energy side is stronger. RIMADIF has the expected high-energy tailing and many counts in the tail and therefore a lower peak.

The shape of the photo peaks in the Cobalt spectrum show a similar shape as for Americium (Figure 8.15 center), but EdgAr+'s peak now reaches 80% of the Ge peak in height. The characteristic lead emission at 72.8, 75.0, 84.9, and 87.3 keV caused by the lead filter is well observed by the Ge detector, Additionally, small lead peaks in the CdTe/CZT spectra originate from lead shields of the experimental setup. It should be mentioned that in subsequent calculations of *PTTotR* the lead peaks in the spectra of all three detectors are replaced by two linear functions, estimating the continuum underneath the peaks to reduce their influence on the calculation. The peak height of the RIMADIF detector is about half of the one of EdgAr+, indicating a lower efficiency of this detector at higher energy. Furthermore, the tail is quite pronounced down to 100 keV when compared to Ge or CdTe, but it hardly exhibits escape peaks and lead peaks are smaller. A more detailed look at the Co-peaks is presented in Figure 8.15 (right).

A statistical comparison on the basis of analyzed spectra for all produced detectors is presented in Chapter 8.7.6.

8.7.4 RIMADIF Thermal Stability Test

Each XD*i* detector's crystal near temperature sensor shall have a target temperature between 20°C and 25°C to limit heat-induced dark current, on the one hand, and to reduce the problem of condensation inside the system, on the other. Its variation over time shall be limited to ±1°C to minimize any gain and offset drifts in the electronics. The upper limit for the acceptable change in peak position with temperature was set to 0.5 keV + 0.1% × E_{Peak} per °C, which is 0.8 resp. 1.1 keV/°C for the 59.5 keV and 122.1 keV line.

In the XD*i* system, the heat generated in the 17 detectors is drawn from them through just one large cooling plate attached from underneath that is flowed through by water circulating through a chiller (Laird MRC300) with 300 W cooling power. Hence, the temperatures of the 17 detector crystals in the system may differ slightly and even fluctuate depending on external temperature. They are monitored to generate an alarm should they leave the tolerance range.

It is of interest to verify the stability of peak position and peak width of each pixel's spectrum, because those that lie on a line making up a detector segment are added together to form a segment spectrum, which is why their calibrations should be optimal. Otherwise, this summation results in a final spectrum with broader and possibly shifted peaks having less similarity with the reference spectra in the detection library.

One method to measure temperature behavior for 17 detectors simultaneously is to investigate it in an XD*i* CBS by irradiating a foam cube that fills the complete object space with X-rays to generate a scatter signal in every relevant pixel of the detector array. The strong tungsten K_α line in each spectrum, originating from the X-ray anode, is used to monitor peak stability while the cooling plate's temperature is varied between 20°C and 25°C in steps of 0.5°C with a measurement done each time. A fit to the resulting average module K_α peak position vs. temperature curve reveals a peak shift gradient of 0.1 keV/°C for RIMADIF detectors, complying with the above-mentioned requirement and indicating that variations of ±1°C at a temperature within 20°C–25°C are acceptable.

8.7.5 EdgAr+ Thermal Stability Test

EdgAr+ has a temperature sensor on each of the array cards in the stack, which allows for detailed temperature monitoring and calculation of an average array card temperature. Throughout the source tests, the cooling plate temperature is held constant at 18.5°C ± 0.1°C, while the mean array card temperature is measured to be 3°C higher with a σ of 0.3°C among the 21 detectors. Looking at the 26 sensors of an EdgAr+ individually reveals a temperature profile along the array card stack with a maximum of 23°C in the middle and a decline by 1.5°C toward the edges.

Temperature stability for EdgAr+ is not measured in an XD*i*-CBS, but just for one detector by using the nuclear sources. In order to quantify the impact of potential temperature changes on energy resolution and peak position, the cooling plate temperature is varied from 16°C to 21°C in steps of 0.25°C to produce array card temperatures between 20°C and 25°C. The detector is illuminated by both sources at each temperature for 600 s. Gain- and offset parameters of all 390 pixels are calculated for the measurement at 22.5°C to establish a reference calibration, which is then applied to all other measurements. Peak position and energy resolution are then determined for all pixel spectra, and respective gradients with temperature are calculated.

The result is that the peak position decreases linearly with increasing temperature at a rate of −0.07 and −0.19 keV/°C for Am and Co, which is sufficiently less than the limits set in the previous chapter.

The energy resolution worsens slightly with increasing temperature at a rate of 0.006 resp. 0.008 keV/°C for the sources. It can therefore be concluded that temperature changes have a negligible impact on peak broadening but have a small impact on peak position, which however would not have a significant effect on diffraction spectra as long as temperature variation can be kept below ±1°C.

8.7.6 Statistical Comparison of RTSD Production-Units and One Germanium Detector

In the following, the definition of dead/bad pixels of RTSD production units is given. Also, spectral parameters as determined in the site acceptance test for 75 RIMADIF, 21 EdgAr+ and one EGPS detector are presented in the form of histograms (Figures 8.16–8.18) and summarized in Table 8.3.

8.7.6.1 Dead and Bad Pixels

In the present evaluation, a pixel is considered "dead", if it has no signal or an energy resolution >7% at 59.5 keV and "bad", if its energy resolution is between 4.5 and 7%. The top of Table 8.3 lists the number of dead and bad pixels for the different detectors. The target values of 3.4% and 11.4% derive

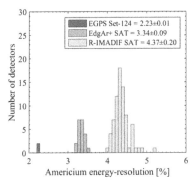 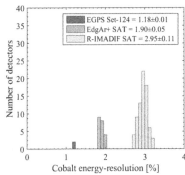

FIGURE 8.16
Histograms showing mean detector energy resolution for EdgAr+ and RIMADIF detectors from small-scale production and the value for EGPS, for [241]Am (left) and [57]Co (right).

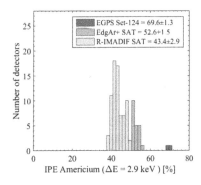 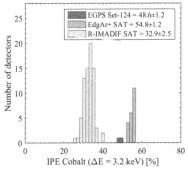

FIGURE 8.17
Histograms showing mean detector intrinsic peak efficiency for EdgAr+ and RIMADIF detectors from small-scale production and the value for EGPS, for [241]Am in a 2.9 keV (left) and for [57]Co in a 3.2 keV peak window (right).

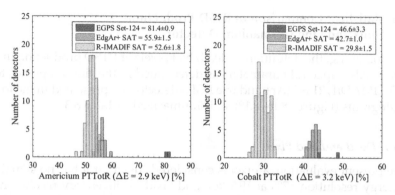

FIGURE 8.18

Histograms showing mean detector peak-to-total ratio for EdgAr+ and RIMADIF detectors from small-scale production and the value for EGPS calculated for ^{241}Am with a 2.9 keV (left) and for ^{57}Co in a 3.2 keV peak window (right).

TABLE 8.3

Mean Performance Characteristics of RIMADIF, EdgAr+, and EGPS Detectors

Mean Detector Characteristic	Unit	Target	RIMADIF Actual	σ	EdgAr+ Actual	σ	EGPS Actual	σ
Detector material			CdZnTe		CdTe		Germanium	
Detectors			75		21		1	
Pixels (segments)/detector			176		390		24	
Dead pixels	%	≤3.4	0.7	0.8	2.2	0.9	0	—
Bad pixels	%	≤11.4	3.5	2.2	0.9	0.5	0	—
ER(^{241}Am)	%	3.0	4.37	0.20	3.34	0.09	2.23	0.01
ER(^{241}Am) variation within detector	%	—	0.33	0.10	0.26	0.06	0.07	0.01
ER(^{57}Co)	%	—	2.95	0.11	1.90	0.05	1.18	0.01
ER(^{57}Co) variation within detector	%	—	0.24	0.04	0.15	0.03	0.15	—
IPE(^{241}Am) (ΔE=4.9%)	%	—	43.4	2.9	52.6	1.5	68.6	3.4
IPE(^{57}Co) (ΔE=2.6%)	%	—	32.9	2.5	54.8	1.2	48.6	1.2
PTTotR Am	%	—	52.6	1.8	55.9	1.5	81.4	0.9
PTTotR Co	%	—	29.8	1.5	42.7	1.0	46.6	3.3

from the requirements that not more than 6 out of 176 pixels of RIMADIF shall be dead and not more than 20 pixels bad. Additionally, there can only be one pixel dead per segment, otherwise the loss in flux is considered unacceptable. The same requirements are chosen for EdgAr+ considering the higher number of pixels though.

The distinction between dead and bad pixels is a compromise to accept detectors with some pixels that have an energy resolution above 4.5%; otherwise, the yield of the CZT head board would have suffered. These pixels degrade the energy resolution of a segment's diffraction spectrum, which would be a reason to exclude them from the line of pixels, but they also contribute to the total flux measured on the respective segment, which is why 20 bad pixels per detector are still accepted.

Table 8.3 shows that with 2.2% dead pixels, EdgAr+ has about thrice as many as RIMADIF but only a fourth as many bad pixels. The variation of the number of bad pixels among detectors is four times higher for RIMADIF. For EGPS, there is no tolerance on dead or bad segments.

8.7.6.2 Energy Resolution Histogram

Figure 8.16 (left) depicts three histograms, each representing the distribution of mean energy resolution at the Americium-line for the two sets of RTSDs produced and the two measurements done with the Ge detector. The latter has an energy resolution of 2.23%, considerably better than EdgAr+ detectors with 3.34%, which however comes respectably close to the desired target of 3%. RIMADIF's energy resolution averages to 4.37% with corrections applied. The standard deviation of the distribution among EdgAr+ is half of that of RIMADIF's (0.1 vs. 0.2% abs.), which indicates a higher potential for repeatedly achieving the same performance in the first case. Looking at the mean variation of energy resolution among the pixels of individual detectors, EdgAr+ has with σ_{ER}= 0.26% also less variation than RIMADIF with 0.33% (see Table 8.3).

In case of the Cobalt histogram, the ratios between the observed energy resolutions above repeat (see Figure 8.16 right) with EGPS at 1.18%, EdgAr+ at 1.90%, and RIMADIF at 2.95%. Again, the standard deviation for RIMADIF with 0.1% is almost twice that of EdgAr+ detectors.

It shall be noted here that the bi-parameter correction of RIMADIF introduces a degradation in energy resolution of ≈0.3% (absolute) at the Americium line, whereas no negative influence at the Cobalt line is observed. The purpose of the correction algorithm is to reduce the tailing by correcting the counts in it, putting them back into the peak. As the algorithm is affected by uncertainties in anode- and cathode-signal, it also adds an uncertainty to the energy of the photon to be corrected, which reflects itself in a broader peak, but with more photons in it. Hence, energy resolution is traded somewhat against higher peak efficiency.

8.7.6.3 Intrinsic Peak-Efficiency Histogram

In Figure 8.17, intrinsic peak efficiencies of the detectors are histogrammed. *IPE* depends strongly on the chosen peak region, which in this case are rather tight windows that certainly exclude some photons of the broader peaks of the RTSDs but allow for a fair comparison with the tight Ge-peaks, which would not profit from broader limits as most counts are already considered.

Both sources exhibit quite high uncertainties in their stated activities, about ±15%, which does not allow to draw conclusions about absolute efficiencies but to compare the detectors to each other.

The average *IPE* at the Americium line for RIMADIF detectors is 43%, for EdgAr+ 53% and 70% for EGPS, based on two measurements, whereby the RIMADIF distribution is twice as broad as EdgAr's.

The data for RIMADIFs was collected over a period of almost 24 months, during which the[57] Co-source's activity dropped in intensity by a factor of six, which may have influenced the background in the spectra. This cannot be the case for EdgAr+, because all data was collected during a week.

Looking at the histograms for the *IPE* at 122 keV, the highest value with 55% comes from the CdTe-detector, followed by Germanium with 49% and CZT with 33%. Surprisingly, EdgAr+ performs better here than EGPS. The reason for this may be the better absorption of CdTe when compared to the 10 mm thick Ge crystal, about 5% more (see Figure 8.10) and the width of the integration window. Although the peak is higher for EGPS, Edgar+'s peak is wider at the rising and falling edge (see Figure 8.15, right) resulting in a larger sum of counts for the window selected.

8.7.6.4 Peak-to-Total Ratio Histograms

The *PTTotR* histogram of Americium (Figure 8.18, left) illustrates that 81% of all counts in a range from 20 to 132 keV end up in a peak window of 2.9 keV for the Ge detector, whereas for EdgAr+ it is 56% and for RIMADIF it is 53%. Obviously, more counts are lost for the RTSDs than for the Ge detector, due to tailing and escaping photons.

For Cobalt (Figure 8.18, right), the CdTe detector performance with a *PTTotR* of 43% comes close to the one of Ge with 44% resp. 49% (1 mm resp. 0.5 mm lead). However, due to the lead filter in front of the source, the Ge-spectrum experiences a higher background, which reduces the *PTTotR* somewhat. Hence, the value of 49% should be interpreted as a lower limit. RIMADIF's *PTTotR* is around 30%. Here the tight peak window of 3.2 keV and the higher tailing are the cause for the difference of 13% when compared to EdgAr+.

8.8 Principle of Threat Detection

As mentioned in Chapter 1.4, about 10 million spectra need to be processed. After summation of pixel to segment spectra, the latter have to be transferred into momentum space under consideration of the scatter angle. Then they can be back projected onto a created voxel grid in 3D object space, where the momentum spectra of different detector focus segment combinations will be added, if they fall into the same voxel. These voxel spectra are the basis for any

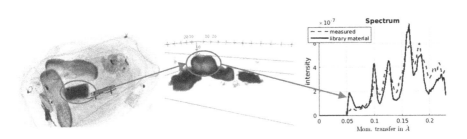

FIGURE 8.19
Left: Normal 2D Image with selected object. *Center*: 3D reconstruction and segmentation of the objects' scatter signal. *Right*: Momentum spectrum of selected object. (Provided by Dr. S. Fritsch.)

further processing, for example, in a first step, the voxel spectra will be integrated to yield the scatter intensity per voxel, from which a 3D scatter intensity map will be generated. Figure 8.19 (left) shows the projection image of a bag and the corresponding scatter intensity distribution (see Figure 8.19, center). A segmentation process will then identify voxels in the map that belong to the same object, whereupon the respective object spectra are built for comparison to spectra of illicit substances in a threat library (see Figure 8.19, right).

8.9 Summary and Conclusion

Considering the seven detector requirements discussed previously, the following can be concluded for CZT and CdTe detectors for XD*i*: (i) spatial resolution is easily achieved through pixelation, (ii) intrinsic peak efficiency suffers from incomplete charge collection, but reaches 43 resp. 53% in a 4.9% window around the Americium peak, and 33 resp. 55% in a 2.6% window around the Cobalt peak. Energy resolution (iii) with 3.3%–4.4% at 60 keV and 1.9%–2.9% at 122 keV is about a factor 1.5 to 2.5 worse than for Ge, but seems acceptable when the overall width of the diffraction peak is considered, (iv) count rate problems do not occur, as the occurring fluxes per pixel channel are far from the limits of the chosen ASICs, (v) detectors are compact without any need for cryogenic cooling and can therefore easily be integrated into an airport environment. Furthermore, RTSDs are (vi) free from maintenance and have a high reliability according to our experience, and (vii) the price per detector area for XD*i*'s RTSDs is less than 20% of the one for the XRD3500's Ge detector and its readout system.

XD*i*-CBS has successfully passed the C1 EU certification for cabin-baggage scanners using RIMADIF detectors and is now going for the next certification level C2. It is expected that future use of EdgAr+ detectors with better energy resolution and efficiency will improve system performance even further.

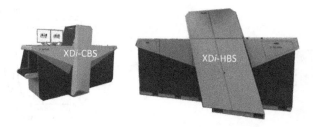

FIGURE 8.20
Photographs of XD*i*-CBS (left) and the XD*i*-HBS prototype (right).

Hence, it can be concluded that RTSDs are one enabling technology for diffraction systems with a large detector area, like XD*i*-CBS for security checkpoints and XD*i*-HBS (see Figure 8.20), which may be tomorrow's false-alarm resolver downstream of other high-speed inspection systems for hold luggage.

Acknowledgments

This work was partially supported by the United States Department of Homeland Security under HSHQDC-11-C-00014.

The author thanks the XD*i* team members for their input and discussions, especially Dr. J.-P. Schlomka and Dr. H. Strecker.

Note

1. Since April 2017 part of Smiths Detection, www.smithsdetection.com.

Bibliography

Awadalla, S. A. et al., 2010. Characterization of detector-grade CdZnTe crystals grown by traveling heater method (THM). *Journal of Crystal Growth*, 312(4), pp. 507–513.
Bearden, J. A., 1967. X-ray wavelengths. *Reviews of Modern Physics*, 39(1), pp. 78–124.
Benshop, T., Mullié, J., Bruins, P. & Martin, J. Y., 2002. Development of a 6 W high reliability cryogenic cooler at Thales Cryogenics. *Proc. SPIE 4820, Infrared Technology and Applications XXVIII*, (23 January 2003). doi: 10.1117/12.451194.

Bolotnikov, A. E. et al., 2008. Effects of Te inclusions on the performance of CdZnTe radiation detectors. *IEEE Transactions On Nuclear Science*, 55(5), pp. 27–57.

Brooks Automation, 2013. *PCC Compact Coolers - Next Generation CryoTiger.* [Online] Available at: www.brooks.com/products/cryopumps-cryochillers/cryochillers/pcc-compact-coolers.

CAEN S.p.A, 2015. *NIM Digitizers N6724.* [Online] Available at: www.caen.it/csite/CaenProd.jsp?idmod=626&parent=12.

Chen, H. et al., 2007. Characterization of traveling heater method (THM) grown Cd0.9Zn0.1Te crystals. *IEEE Transactions on Nuclear Science*, 54(4), pp. 811–816.

Cozzini, C. et al., 2010. Energy dispersive X-ray diffraction spectral resolution considerations for security screening applications. Knoxville, *IEEE Nuclear Science Symposium.*

ECAC, 2017a. *ECAC-CEP-EDSCB-Web-Update-20-September-2017.pdf.* [Online] Available at: www.ecac-ceac.org/cep.

ECAC, 2017b. *ECAC-CEP-EDS-Web-Update-12-October-2017.pdf.* [Online] Available at: www.ecac-ceac.org/cep.

Funaki, M. et al., 2007. *Development of CdTe in Acrorad.* [Online] Available at: www.acrorad.co.jp/dcms_media/other/Development%20of%20CdTe%20detectors%20in%20Acrorad.pdf.

Gevin, O. et al., 2012. Imaging X-ray detector front-end with high dynamic range: IDeF-X HD. *Nuclear Instruments and Methods in Physics Research Section A: Accelerators, Spectrometers, Detectors and Associated Equipment*, 695, pp. 415–419.

Gostilo, V. et al., 2002. The development of drift-strip detectors based on CdZnTe. *IEEE Transactions on Nuclear Science*, 49(5). doi:10.1109/NSSMIC.2001.1009278.

Harding, G., 1989. X-ray quanta measuring device including diaphragm for producing conical radiation beam on object being measured. US, Patent No. 5,008,911.

Harding, G., 2005. The design of direct tomographic energy-dispersive X-ray diffraction imaging (XDi) systems. *Proc. SPIE 5923, Penetrating Radiation Systems and Applications VII*, Volume 59230R.

Harding, G., 2009. Compact multi-focus X-ray source, X-ray diffraction imaging system, and method for fabricating compact multi-focus X-ray source. US, Patent No. 7,756,249 B1.

Harding, G., 2009. X-ray diffraction imaging-a multi-generational perspective. *Applied Radiation and Isotopes*, 67(2), pp. 287–295.

Harding, G. & Delfs, J., 2007. Liquids identification with x-ray diffraction. *Proc. SPIE 6707*, p. 67070T.

Harding, G. et al., 2012. X-ray diffraction imaging with the multiple inverse fan beam topology: Principles, performance and potential for security screening. *Applied Radiation and Isotopes*, 70(7), pp. 1228–1237.

Harding, G., Kosciesza, D., Olesinski, S. & Strecker, H., 2010. Experimental comparison of Next-Generation XDi topologies. *Proc. SPIE 7806, Penetrating Radiation Systems and Applications XI*, Volume 78060I.

Harding, G., Strecker, H. R. & Delfs, J., 2008. Primary collimator and systems for X-ray diffraction imaging, and method for fabricating a primary collimator. United States, Patent No. US20100135462 A1.

Harding, G., Strecker, H., Kosciesza, D. & Gordon, J., 2009. Detector considerations relevant to X-ray diffraction imaging for security screening applications. *Proc. SPIE 7306*, p. 730619.

Hartick, M., 2017. Head of Development Department, Smiths Heimann GmbH [Interview] (23 Nov. 2017).

He, Z., 2001. Review of the Shockley–Ramo theorem and its application in semiconductor gamma-ray detectors. *Nuclear Instruments and Methods in Physics Research A*, 463, pp. 250–267.

IDEAS, 2017. *IDEAS-IC-Products / Readout ICs for X-rays, Gamma-rays, and Elementary Particles*. [Online] Available at: http://ideas.no/ideas-ic-products/[Accessed 1 9 2017].

Iwanczyk, J. S., Nygard, E. & Kosciesza, D., 2014. Cadmium telluride detector system for identifying illicit substances by X-ray energy dispersive diffraction. *Nuclear Science Symposium and Medical Imaging Conference (NSS/MIC), IEEE*.

Iwanczyk, J. S., Nygard, E. & Saveliev, V. D., 2011. *Edge-on two-dimensional detector arrays*. US, Patent No. 8,513,617.

Iwanczyk, J., Szymczyk, W. M., Traczyk, M. & Triboulet, R., 1979. X-ray fluorescence escape peaks in gamma-ray spectra detected by CdTe detectors. *Nuclear Instruments and Methods*, 165, pp. 289–295.

Knoll, G. F., 1989. *Radiation Detection and Measurement*. 2nd ed. New York: John Wiley & Sons, 117 ff.

Kuvvetli, I. & Budtz-Jorgensen, C., 2005. Pixelated CdZnTe drift detectors. *IEEE Transactions on Nuclear Science*, 52(5), pp. 1975–1981.

Luggar, R. D. et al., 1995. Optimization of a low angle X-ray scatter system for explosive detection. *SPIE Law Enforcement Technologies*, 2511, p. 46.

Montemont, G. et al., 2007. Simulation and design of orthogonal capacitive strip CdZnTe detectors. *IEEE Transactions on Nuclear Science*, 54, pp. 854–859.

Montémont, G. et al., 2013. An Autonomous CZT Module for X-ray Diffraction Imaging. *Nuclear Science Symposium and Medical Imaging Conference (NSS/MIC), IEEE*.

Morpho Detection, 2012. *System of Systems*. [Online] Available at: www.morpho.com/sites/morpho/files/morpho_system_of_systems_dat_r001_0.pdf.

Morpho Detection, 2015. *Technical Specifications XRD3500*. [Online] Available at: www.morpho.com/sites/morpho/files/morpho_xrd3500_ts_r002_0.pdf.

Niemelä, A., Sipilä, H. & Ivanov, V. I., 1996. Improving CdZnTe X-ray detector performance by cooling and rise time discrimination. *Nuclear Instruments and Methods in Physics Research Section A*, 377(2–3), pp. 484–486.

Procz, S. et al., 2013. Medipix3 CT for material sciences. *Journal of Instrumentation*, 8. doi:10.1088/1748-0221/8/01/C01025.

Ries, H., 2003. Multi level concept for the detection of explosives. In: H. Schubert & A. Kuznetsov, eds. *Detection of Bulk Explosives*. St. Petersburg: Kluwer Academic Publishers, p. 45 ff.

Ries, H. e. a., 2001. Apparatus for determining the crystalline and polycrystalline materials of an item. US, Patent No. 6,542,578.

Rostaing, J.-P., 2017. *A low-noise 16-Channel Charge Amplifier IC for X-ray Spectrometry*. [Online] Available at: http://cmp3.imag.fr/aboutus/gallery/details.php?id_circ=474&y=2013.

Rumaiz, A. K. et al., 2014. A monolithic segmented germanium detector with highly integrated readout. *IEEE Transactions on Nuclear Science*, 61(6), pp. 3721–3726.

Sanchez del Rio, M. & Dejus, R. J., 2004. Status of XOP: An X-ray optics software toolkit. *SPIE Proceedings*, 5536, pp. 171–174.

Seino, T. & Takahashi, I., 2007. CdTe detector characteristics at 30°C and 35°C when using the periodic bias reset technique. *IEEE Transactions on Nuclear Science,* 54(4), pp. 777–781.

Shih, A. Y. et al., 2012. The Gamma-Ray Imager/Polarimeter for Solar flares (GRIPS). *Space Telescopes and Instrumentation 2012 (SPIE): Ultraviolet to Gamma Ray, 8443(4H).*

Shiraki, H. et al., 2007. Improvement of the productivity in the growth of CdTe single crystal by THM for the new PET system. *Nuclear Science Symposium Conference Record.*

Smiths-Heimann, 2004. *Solutions for a safer world - HI-SCAN 10065 HDX.* [Online] Available at: www.donggok.co.kr/data/HS10065HDX.pdf

Stein, J., 2017. *target Systemelektronik.* [Online] Available at: http://target-sg.com/.

Strecker, H. et al., 1993. Detection of explosives in airport baggage using coherent X-ray scatter. *SPIE Substance Detection Systems,* 2092, 399.

Szeles, C. & Driver, M. C., 1998. Growth and properties of semi-insulating CdZnTe for radiation detector applications. *SPIE Conference on Hard X-ray and Gamma-Ray Detector Physics and Applications,* 3446(1).

Takahashi, T. et al., 1998. Performance of a new Schottky CdTe detector for hard X-ray spectroscopy. *IEEE Transactions on Nuclear Science,* 45(3), pp. 428–432.

Takahashi, T. et al., 2001. High resolution CdTe detector and applications to imaging devices. *IEEE Transactions on Nuclear Science,* 48(3), pp. 287–291.

Takahashi, T. et al., 2002. High-resolution Schottky CdTe diode detector. *IEEE Transactions on Nuclear Science,* 49(3), pp. 1297–1303.

Thales Cryogenics, 2017. *LPT9310.* [Online] Available at: www.thales-cryogenics. com/products/coolers/lpt/lpt9310/.

Verger, L. et al., 2004. Performance and perspectives of a CdZnTe-based gamma camera for medical imaging. *IEEE Transactions on Nuclear Science,* 51(6), pp. 3111–3117.

Warren, B. E., 1990. *X-Ray Diffraction.* 1st ed. New York: Dover Publications.

Index